# ADVANCED BUILDING MEASUREMENT

**Macmillan Building and Surveying Series**
Series Editor: Ivor H. Seeley
                Emeritus Professor, Nottingham Polytechnic
*Advanced Building Measurement, second edition*   Ivor H. Seeley
*Advanced Valuation*   Diane Butler and David Richmond
*An Introduction to Building Services*   Christopher A. Howard
*Applied Valuation*   Diane Butler
*Asset Valuation*   Michael Rayner
*Building Economics, third edition*   Ivor H. Seeley
*Building Maintenance, second edition*   Ivor H. Seeley
*Building Procurement*   Alan Turner
*Building Quantities Explained, fourth edition*   Ivor H. Seeley
*Building Surveys, Reports and Dilapidations*   Ivor H. Seeley
*Building Technology, third edition*   Ivor H. Seeley
*Civil Engineering Contract Administration and Control*
   Ivor H. Seeley
*Civil Engineering Quantities, fourth edition*   Ivor H. Seeley
*Civil Engineering Specification, second edition*   Ivor H. Seeley
*Computers and Quantity Surveyors*   Adrian Smith
*Contract Planning and Contractual Procedures*   B. Cooke
*Contract Planning Case Studies*   B. Cooke
*Environmental Science in Building, second edition*   R. McMullan
*Housing Associations*   Helen Cope
*Introduction to Valuation*   D. Richmond
*Principles of Property Investment and Pricing*   W.D. Fraser
*Quality Assurance in Building*   Alan Griffith
*Quantity Surveying Practice*   Ivor H. Seeley
*Structural Detailing*   P. Newton
*Urban Land Economics and Public Policy, fourth edition*
   P.N. Balchin, J.L. Kieve and G.H. Bull
*Urban Renewal*   Chris Couch
*1980 JCT Standard Form of Building Contract, second edition*
   R.F. Fellows

**Other titles by the same authors**
*Housing Improvement and Social Inequality*   P.N. Balchin (Gower)
*Housing Policy: An Introduction*   P.N. Balchin (Croom Helm)
*Housing Policy and Housing Needs*   P.N. Balchin (Macmillan)
*Regional and Urban Economics*   P.N. Balchin and G.H. Bull
   (Harper and Row)
*The Electric Telegraph: An Economic and Social History*   J.L. Kieve
   (David and Charles)

# ADVANCED BUILDING MEASUREMENT

**IVOR H. SEELEY**
BSc(Est Man), MA, PhD, FRICS,
CEng, FICE, FCIOB, MIH

*Chartered Quantity Surveyor*
*Emeritus Professor of Nottingham Polytechnic*

Second Edition

MACMILLAN

© Ivor H. Seeley 1989

All rights reserved. No reproduction, copy or transmission
of this publication may be made without written permission.

No paragraph of this publication may be reproduced, copied or
transmitted save with written permission or in accordance with
the provisions of the Copyright, Designs and Patents Act 1988,
or under the terms of any licence permitting limited copying
issued by the Copyright Licensing Agency, 33–4 Alfred Place,
London WC1E 7DP.

Any person who does any unauthorised act in relation to
this publication may be liable to criminal prosecution and
civil claims for damages.

First edition 1982
Reprinted 1985, 1986
Second edition 1989
Reprinted 1990

Published by
MACMILLAN EDUCATION LTD
Houndmills, Basingstoke, Hampshire RG21 2XS
and London
Companies and representatives
throughout the world

Printed in Hong Kong

British Library Cataloguing in Publication Data
Seeley, Ivor H. (Ivor Hugh), *1924–*
    Advanced building measurement. – 2nd ed
    1. Great Britain. Buildings. Construction.
    Standard method of measurement
    I. Title
    692'.3

    ISBN 0–333–48542–4
    ISBN 0–333–48543–2   Pbk

'The only place where
success comes before work is in a dictionary'
Vidal Sassoon

This book is dedicated to those numerous quantity surveyors and quantity surveying graduates and students in the United Kingdom and from other countries who over the years have given me the benefit of their ideas and views and who have, in so doing, contributed to the preparation of this book.

# CONTENTS

*Measurement examples* ix

*Preface* xi

*Acknowledgements* xiii

1 PRELIMINARIES, PREAMBLES, CO-ORDINATED PROJECT INFORMATION AND USE OF COMPUTERS IN BILL PRODUCTION  1

General introduction; Scope of Preliminaries Bill and introductory information; Preliminary information; Contractual aspects; Employer's requirements; Contractor's general costs; Works by nominated sub-contractors, goods and materials from nominated suppliers and works by statutory authorities; Prime cost and provisional sums and direct works; Dayworks; General summary; Preambles; Typical preamble clause headings; Specimen preamble clauses — Excavating and filling; Structural steel framing; Co-ordinated project information; Use of computers in bill production.

2 SUBSTRUCTURES  26

Measurement procedures; Excavating and filling — generally: Site preparation; Excavating; Earthwork support; Disposal of water and excavated material; Filling; Surface treatments; Sloping site excavation and associated works; Piling; Underpinning; Worked examples.

3 BRICKWORK, BLOCKWORK AND MASONRY  47

Brickwork and blockwork — General requirements; General brickwork and blockwork; Brick and block facework; Other classes of work; Damp-proof courses and sundries to brickwork and blockwork; Masonry — measurement requirements; Description of masonry; Measurement procedures; Rubble walling; Worked examples.

4 REINFORCED CONCRETE  63

Categories; Measurement of reinforced concrete; *In situ* concrete; Reinforcement; Formwork; Sequence of measurement of reinforced concrete floors and roofs; Measurement of reinforced concrete staircase; Precast concrete; Precast concrete/composite decking; Pre-stressed concrete work; Worked examples.

5 STRUCTURAL STEELWORK  83

General introduction; Principles of measurement of structural steel framing; Fabrication of framing; Fittings, connections and special bolts; Surface preparation and treatment; Measurement procedure for structural steel framing; Worked example.

6 WOODWORK AND METALWORK  91

Timber particulars; Carpentry/timber framing; Flooring and roof boarding; Linings and casings; Composite items — Fitments; Doors and adjoining screens; Windows; Glazed screens; Staircases; Worked examples.

7 FINISHINGS **107**

Sequence of measurement; General principles of measurement — Ceilings; Walls; Floors; Other finishings; Skirtings and similar members; Painting and decorating; Worked example.

8 MECHANICAL SERVICES **121**

General background; Measurement procedures; Approach to measurement; Other measurement aspects; Worked example.

9 ELECTRICAL SERVICES **135**

General background; Measurement procedures; Detailed measurement; Worked example.

10 ALTERATION WORK AND REVISION NOTES **146**

Extent of work; Measurement of demolition, alterations and repair work — Demolishing structures/shoring; Alterations; Repairing and renovating concrete, brick, block and stone; Sequence of measurement; Worked example; Revision notes — General aspects; Weaknesses in specific work sections.

*Bibliography* 162

*Index* 163

# MEASUREMENT EXAMPLES

(1) Stepped foundations
(2) Piling (bored cast in place concrete piles, preformed concrete piles and interlocking steel piles)
(3) Underpinning
(4) Brick chimney shaft
(5) Masonry (stone surround to door opening, stone balustrade and plain ashlar facing)
(6) Reinforced concrete suspended floors
(7) Reinforced concrete staircase and balcony
(8) Prestressed precast concrete bridge
(9) Structural steelwork
(10) Doors and adjoining lights
(11) Fixed benching
(12) Finishings
(13) Mechanical services
(14) Electrical services
(15) Alterations

# PREFACE

This book provides a logical extension to the worked examples contained in *Building Quantities Explained*. It is intended to assist students in the later years of degree and higher diploma and certificate courses in quantity surveying and building, and those studying for the later examinations of the appropriate professional bodies, such as RICS, CIOB, IAAS and ASI.

The book was titled *Advanced Building Measurement*, rather than Advanced Quantities, to match the appropriate subject heading of most of the relevant examinations. The main objective has been to produce a good selection of worked examples, supported by comprehensive explanatory notes, and covering a reasonable range of constructional components that the candidate may face in the examination. It will be appreciated that however large the book it could not cover every conceivable type of work but it is hoped that the range selected will prove reasonably comprehensive and helpful to the student. In the first instance it was the intention to produce a loose-leaf book with large folded pull-out drawings but the high cost of production prevented its implementation.

The measurement examples in the second edition follow closely the principles outlined in SMM7 and the Code of Procedure for Measurement of Building Works. Particular attention has been paid to securing accuracy in dimensions, ample use of waste calculations and locational notes, adequate descriptions, a logical sequence of items and a good standard of presentation, all of which are important criteria in which students should endeavour to achieve proficiency. A number of dimensioned and bill diagrams have been included to illustrate their nature and use. To make the examples more realistic some take off and query lists have been incorporated.

Apart from describing the measurement procedures and providing worked examples, a chapter has also been included on Preliminaries and Preambles, as examination questions are set from time to time on these particular aspects. Their format and content does also influence the billed descriptions of the measured items. Bill formats and bill preparation processes have not been included as it is felt that they have been covered sufficiently in *Building Quantities Explained*. However, it was felt desirable to outline the documents and procedures encompassed in co-ordinated project information, and the use of computers in bill production. Some revision notes based on examiners' reports have been included at the end of the text in chapter 10, as these identify major weaknesses and will assist students with their revision work.

Nottingham  
Spring 1989

IVOR H. SEELEY

# ACKNOWLEDGEMENTS

The author expresses his thanks to the Standing Joint Committee for the Standard Method of Measurement of Building Works for kind permission to quote from *SMM7: Standard Method of Measurement of Building Works: Seventh Edition* and the *SMM7 Code of Procedure for Measurement of Building Works*. Considerable help and advice was received from Christopher Willis FRICS FCIArb and Norman Wheatley FRICS, Chairman and Honorary Secretary respectively of the Standing Joint Committee. I am also grateful to the Society of Chief Quantity Surveyors in Local Government for permission to quote extensively from its report on *The Presentation and Format of Standard Preliminaries*, and to the former Greater London Council and to the Architectural Press for permission to quote from *Preambles to Bills of Quantities*. Adrian Smith kindly gave permission to quote and incorporate a figure from *Computers and Quantity Surveyors*, in the same series.

Considerable assistance has been obtained from a variety of persons and organisations, many of whom are listed below, which the author acknowledges with gratitude. The Royal Institution of Chartered Surveyors, former Institute of Quantity Surveyors, Polytechnic of Wales (Department of Estate Management and Quantity Surveying), Polytechnic of the South Bank (Department of Building Economics), Trent Polytechnic (Department of Surveying) and University of Hong Kong (School of Architecture) kindly agreed to the use of past examination paper drawings which have been incorporated to advantage, generally with a number of adaptations to increase their usefulness to the student. Subject to various modifications, some of the examples follow concepts adopted by the author when preparing study material for the Ellis School of Surveying and Building in Worcester in years past. This wide ranging search for suitable material for examples has enabled the author to provide a good coverage of the more commonly used works sections in SMM7, without repetition of similar items, and of an appropriate standard and format.

Keith Stafford FRICS helped with the preparation of the original draft material for chapters 8 and 9. Jack Preite provided various drawings of alteration work from which drawing 15 was prepared and E. G. Phillips, Son and Partners, Consulting Engineers, Nottingham assisted with the supply of a drawing and specification that provided the basis for the mechanical services example. Nigel Neville FRICS, while serving as Regional Quantity Surveyor to the East Anglian Regional Health Authority, kindly supplied some excellent documents produced in his office and illustrating current practice.

Ronald Sears MCIOB once again prepared the handwritten dimensions and explanatory notes. He spares neither time nor effort in producing artwork of outstanding quality which adds immeasurably to the value of the book. Malcolm Stewart of Macmillan Education Ltd resolved all the book production problems in a most satisfactory and helpful way. I am grateful to the Earl of Stockton and Peter Murby of the publishers for their general encouragement, and to my wife for her continual tolerance and understanding.

---

*Publisher's note:* Pages 31, 43, 51, 79, 95, 101 and 127 are intentionally blank to improve presentation of the drawings and dimensions sheets.

# 1 PRELIMINARIES, PREAMBLES, CO-ORDINATED PROJECT INFORMATION AND USE OF COMPUTERS IN BILL PRODUCTION

## GENERAL INTRODUCTION

One professional examination examiner's report stated that many candidates were unable to distinguish between preliminary items and preambles, even when a copy of the Standard Method was available to them.

A *Preliminaries Bill* is normally the first bill in a bill of quantities and its main purpose is to set out all the general liabilities and obligations of the Contractor covering the contract in its entirety, in order that the Contractor may have the opportunity to price them individually, rather than to have to include them in the billed rates for specific items of work. *Preamble clauses* are usually incorporated at the head of each work sectional bill, covering matters that relate to the particular work section and affect the price for the work. Preamble clauses often contain materials and workmanship requirements where no project specification is prepared.

With both types of documentation, it is important to adopt a systematic and logical approach and to ensure that all appropriate matters relevant to the specific contract are included. The normal procedure of using previous contract documents as a guide can give rise to problems where there are extensive variations in the scope and nature of the two contracts.

## SCOPE OF PRELIMINARIES BILL AND INTRODUCTORY INFORMATION

The contents of a Preliminaries Bill are usually based on the particulars listed in Section A (Preliminaries/General Conditions) of SMM7. These particulars include details of the conditions of contract, clause headings from the conditions and appropriate insertions to the conditions (SMM A20). In addition further clauses may be needed to describe special conditions or requirements peculiar to the specific contract. The Preliminaries clauses of SMM7 are examined and examples given where it is considered helpful to illustrate a suitable approach or to clarify particular points.

The Society of Chief Quantity Surveyors in Local Government (SCQSLG), in its excellent report 'The Presentation and Format of Standard Preliminaries for use with JCT Standard Form of Building Contract with Quantities 1980 Edition' (1981), recommends that the Preliminaries Bill should be preceded by particulars of procedures of which tenderers need to be informed and which do not relate to any matters contained in Section B of SMM6 (now Section A of SMM7). Some examples of the type of information and instructions given to tenderers follow, based on procedures commonly adopted in local government offices.

(1) The Employer shall observe the general principles contained in the Code of Procedure for Single Stage Selective Tendering 1977, published by the NJCC for Building.

(2) The bill of quantities comprises pages in numerical sequence and the Contractor shall check the number of each page and if any are missing or duplicated or if any word, letter or figure is indistinct, he shall notify the Quantity Surveyor.

(3) No alteration shall be made in the text of the bill of quantities unless expressly instructed in writing by the Quantity Surveyor before the tender is submitted.

(4) One copy of the bill of quantities priced and extended in ink shall be returned to the Architect/Supervising Officer/Quantity Surveyor in the envelope provided by not later than the time for submission of the tender. Alternatively a Contractor whose tender is under consideration may be required to submit this document within . . . working days of being requested to do so.

(5) The priced bill of quantities will be examined by the Quantity Surveyor to detect errors in the computation of the tender. In the event of the discovery of any errors the procedure described in Section 6.3, alternative 1, of the Code of Procedure for Single Stage Selective Tendering 1977 will be adopted.

(6) Tenders shall remain open for consideration for a period of two months from the date for receipt, after which they shall be subject to confirmation.

Readers are referred to the SCQSLG publication on Standard Preliminaries described on p. 1 for comprehensive guidance on the preparation of Preliminaries.

## PRELIMINARY INFORMATION

SMM A10 prescribes that the name, nature and location of the project and the names and addresses of the Employer and Consultants shall be listed. Some surveyors consider it unnecessary to give the name of the project as it will appear on the cover of the bill of quantities, although the author does not consider this limited duplication to be a matter of real concern.

A list of drawings from which the bill of quantities was prepared is included (SMM A11) and details of the site, such as site boundaries, adjoining buildings and existing mains and services (SMM A12). SMM A13 provides for the inclusion of a description of the work, covering the elements of each new building, with the dimensions and shape of each building, and details of related work by others.

Some typical bill entries follow to illustrate a common approach, and they would in practice be itemised and printed on billed paper with pricing columns. The total sum entered by the Contractor at the bottom of each page of the Preliminaries Bill will be transferred to a Collection at the end of the Bill.

### 1.00 Preliminary Particulars

*Project Particulars* (as SMM A10.1.1–2.0.0)
1.01 *Name and nature.* The Newtown Health Centre project comprises the erection of a single storey health centre, containing general practice consulting suites and other primary care facilities.
1.02 *Location.* The site is situated in Black Street, Newtown, Cambridgeshire.
1.03 *Employer and consultants.* Details will be inserted of the names and addresses of the Employer, Architect and Quantity Surveyor.
1.04 *Other consultants.* Details of the names and addresses of other consultants not named in the contract conditions are inserted, such as the Services Engineer and the Structural Engineer.

SMM A11 requires a list of the drawings from which the bill of quantities has been prepared and which will be available for inspection by the Contractor. The drawings may fall into several categories.

*Drawings*
1.05 *Bill preparation drawings.* The following drawings were used for the preparation of the bill of quantities (as SMM A11.1.1.0.0.)

    *Drawing Nr*                 *Scale/Description*

1.06 *Tender drawings.* The following drawings accompany the bill of quantities

    *Drawing Nr*                 *Scale/Description*

1.07 *Contract drawings.* The following drawings form part of the contract

    *Drawing Nr*                 *Scale/Description*

1.08 *Inspection of documents.* A copy of the bill preparation drawings which do not accompany the bill of quantities may be inspected at .......... during usual office hours (by prior appointment telephone ............).

SMM A12 requires the inclusion of site conditions like the particulars that follow.

*Description of the Site* (as SMM A12.1.1–4.0.0)
1.09 *Boundaries.* The boundaries of the site are indicated on drawing HC 102/1.
1.10 *Access.* Access to the site is directly off Black Street by means of a temporary crossover.
1.11 *Existing services.* Services that are known to exist on or over the site are shown on drawing HC 102/1.
1.12 *Abutting buildings.* The Contractor's attention is drawn to the old stone cottage abutting the eastern boundary of the site.
1.13 *Visit to site.* The tenderer is recommended to visit the site and to ascertain all local conditions and restrictions likely to affect the works.

The arrangements for visiting the site are
.................
A copy of the soil investigation report is incorporated elsewhere in the bill of quantities.

The description of the work follows the rules contained in SMM A13.1.1–3.1–3.0. The Code of Procedure emphasises that its purpose is to give the estimator an initial impression of the type of work involved. Much of the information will be readily available from the drawings that accompany the tender documents and further general information can be obtained from the descriptions which precede individual sections in the bill of quantities. Hence the initial description can be relatively succinct, as illustrated in the following example.

*Description of the Work*
1.14 *Description of the work.* The work comprises the erection of a single storey health centre, overall size 36.50 × 17.20 × 2.26 m high to eaves, with a total floor area of 542 m². The construction consists of concrete trench fill foundations, external cavity walls of faced brick and concrete block with aluminium windows and doors in hardwood frames. The roof is of concrete covered with asphalt. Internally, walls are plastered and painted.

Services comprise low pressure hot water and heating, electrical installation, fire alarms and emergency lighting. External works include services, drainage, fences, car park and landscape work.

## CONTRACTUAL ASPECTS

SMM A20.1.1.–5.0.1 covers the Form of Contract and it needs to be stated whether the contract is under seal. Normally, standard conditions of contract are used, when particulars of the edition to be used and a schedule of clause headings are to be given. Where there is provision for alternative or optional clauses, the operative clauses are to be stated. Amendments to standard clauses and special clauses are given in full. Where there is an appendix to the conditions of contract requiring insertions, a schedule of insertions is set out in the bill. The Employer's insurance responsibility and the nature of the performance guarantee bond are to be stated where applicable. A typical approach adopted by many local government quantity surveyors, based mainly on the SCQSLG document, follows; minor changes would be needed for use with the private edition of the JCT Standard Form of Building Contract.

**2.00 Contract**

*Form, Type and Conditions of Contract*
2.01 *Contract under seal.* The tenderer's attention is drawn to the fact that the main contract will be under seal.
2.02 *Form of contract.* The form of contract is the JCT Standard Form of Building Contract, Local Authorities with Quantities, 1980 Edition.
2.03 *Schedule of clause headings.* Clause numbers and headings of the Schedule of Conditions of the Standard Form of Building Contract:

*Schedule of clause headings*
1 Interpretation and definitions
2 Contractor's obligations
3 Contract Sum — additions or deductions — adjustment — interim certificates
4 Architect's/Supervising Officer's instructions
5 Contract documents — other documents — issue of certificates
6 Statutory obligations, notices, fees and charges (items for rates on temporary buildings and water for the Works are provided elsewhere)
7 Levels and setting out of the Works
8 Materials, goods and workmanship to conform to description, testing and inspection
9 Royalties and patent rights
10 Person-in-charge (see item 4.01)
11 Access for Architect/Supervising Officer to the Works
12 Clerk of Works
13 Variations and provisional sums
14 Contract Sum
15 Value added tax — supplemental provisions
16 Materials and goods unfixed or off site
17 Practical completion and defects liability (see Appendix)
18 Partial possession by Employer (see Appendix)
19 Assignment and Sub-Contracts
19A Fair wages
20 Injury to persons and property and Employer's indemnity
21 Insurance against injury to persons and property (see Appendix and General Summary) (provisional sum required where insurable risk considered to exist)
22 Insurance of the Works against clause 22 perils
Clause 22A shall apply (see Appendix and General Summary) (need for adequate insurance cover by Contractor)
23 Date of possession, completion and postponement (see Appendix)
24 Damages for non-completion (see Appendix)
25 Extension of time
26 Loss and expense caused by matters materially affecting regular progress of the Works
27 Determination by Employer
28 Determination by Contractor (see Appendix)
29 Works by Employer or persons employed or engaged by Employer

30 Certificates and payments (see Appendix)
31 Finance (No.2) Act 1975 — Statutory tax deduction scheme (Contractor's tax certificate to be inspected and Contractor to satisfy himself as to exemption status of all sub-contractors)
32 Outbreak of hostilities
33 War damage
34 Antiquities
35 Nominated Sub-contractors (all nominated sub-contracts under seal and all Nominated Sub-contractors bound to Employer under terms of JCT Standard Form of Employer/Nominated Sub-contractor Agreement)
36 Nominated Suppliers
37 Fluctuations (see Appendix)
Clause 38, 39 or 40 shall apply.

2.04 *Appendix to Conditions Clause*

| | | |
|---|---|---|
| Statutory tax deduction scheme Finance (No. 2) Act 1975 | Fourth recital and 31 | |
| Settlement of disputes —arbitration | 5.1 | |
| Date for completion | 1.3 | 30 November 1990 |
| Defects Liability Period | 17.2 | 6 months from day named in Certificate of Practical Completion of the Works, except for heating installation for which the period shall be 12 months |
| Insurance cover for any one occurrence or series of occurrences arising out of one event | 21.1.1 | £3 000 000 |
| Percentage to cover professional fees | 22A | 12% |
| Date of possession | 23.1 | 27 November 1989 |
| Liquidated and ascertained damages | 24.2 | at the rate of £600 per week |
| Period of delay | 28.1.3 | |
| (i) by reason of loss or damage caused by any one of the Clause 22 perils | | 3 months |
| (ii) for any other reason | 28.1.3.1, 28.1.3.3 to 28.1.3.7 | 1 month |
| Period of Interim Certificates | 30.1.3 | 1 month |
| Retention percentage | 30.4.1.1 | 3% |
| Period of Final Measurement and Valuation | 30.6.1.2 | 12 months from the day named in the Certificate of Practical Completion of the Works |
| Period for issue of Final Certificate | 30.8 | 3 months |
| Work reserved for Nominated Sub-contractors for which the Contractor desires to tender | 35.2 | |
| Fluctuations (if alternative is not shown clause 38 shall apply) | 37 | clause 38 clause 39 clause 40 |
| Percentage addition | 38.7 or 39.8 | 10% |
| Formula Rules | 40.1.1.1 | |
| rule 3 | | Base Month October 1989 |
| rule 3 | | Non-Adjustable Element 10% |
| rules 10 and 30.1 | | Part I of Section 2 of the Formula Rules is to apply |

*Contractor's Liability*
2.05 *Damage and theft.* The Contractor shall be solely responsible for safeguarding the Works, materials and plant against damage and theft and he is advised to effect such additional insurance as is necessary adequately to cover such risks.

*Employer's Liability*
2.06 *Insurance.* See note under item 2.03 (clause 21).

*Local Authorities Fees and Charges*
2.07 *Provisional Sum.* A provisional sum for fees and charges, including rates upon temporary buildings which the Contractor is required to pay to local authorities, is included elsewhere in the bill of quantities.

## EMPLOYER'S REQUIREMENTS

A summary of likely items is listed in SMM A30 to A37 and they form a useful check list. Provision is normally made in the bill of quantities for the insertion of fixed and/or time related charges by the Contractor against each individual item. Typical entries follow and the appropriate SMM7 reference is inserted against each group of items for the benefit of the reader.

### 3.00 Employer's Requirements

*Tendering/Sub-letting/Supply* (as SMM A30.1.1.1.–2.0)
3.01 *Sufficiency of tender.* The Contractor shall be deemed to have satisfied himself before tendering as to the correctness and sufficiency of his tender for the works and of the prices inserted in the bill of quantities, to cover all his obligations under the contract and everything necessary for the proper completion of the works.

3.02 *Sub-letting.* The Contractor shall take note that certain work measured or otherwise described in the Bill and priced by the Contractor must be carried out by persons named in a list contained in the Bill.

3.03 *Supply of materials and goods.* The Contractor shall ensure that all necessary materials and goods shall be delivered to the site in the correct quantities and quality and at the required time.

*Provision, Content and Use of Documents*
(as SMM A31.1.1.1–2.0)

3.04 *Upkeep of documents on the site.* All loose copies of documents which are required to be maintained by the Contractor on the site, such as one reference set of drawings with any revisions, site instructions and delivery notes, shall be kept fastened in suitable covers/files in the site office, for protection against loss or damage and for easy accessibility. Any document removed from these files shall be recorded therein so that it can be traced when required. At practical completion the Contractor shall request the Architect's instructions regarding the further treatment of the collected documents.

3.05 *Confidentiality of documents.* The Contractor shall treat the contract and everything within it as private and confidential. In particular, the Contractor shall not publish any information, drawing or photograph relating to the works, except with the written consent of the Architect and subject to such conditions as he may prescribe.

*Management of the Works* (as SMM A32.1.1.1–2.0)

3.06 *Suspension of works during bad weather.* The Contractor shall, without compensation, delay or suspend the progress of the works or any part thereof, during frost or bad weather for such periods as may be required by the Architect. The Architect shall determine what extension of time (if any) shall be allowed to the Contractor for such suspensions.

*Quality Standards/Control* (as SMM A33.1.1.1–2.0)

3.07 *Materials and workmanship.* Materials, components and workmanship shall be of good quality and in accordance with the British Standards, Agrément Certificates and Codes of Practice prescribed.

3.08 *Quality control.* The Contractor shall carry out the various quality control procedures described in the specification to ensure that the required quality standards are achieved.

3.09 *Tests.* The Architect may, when he considers it advisable, test any materials before they leave the maker's premises as well as after delivery on site, and the Architect shall be at liberty to reject any materials after delivery should he consider them unsatisfactory, notwithstanding the preliminary test and approval of the materials at the maker's premises. The costs of these tests are to be borne by the Contractor.

*Security/Safety/Protection* (as SMM A34.1.1–7.1–2.0)

3.10 *Noise and pollution control.* The Contractor's attention is drawn to Sections 60 and 61 of the Control of Pollution Act 1974 with reference to the control of noise in relation to any demolition and construction works and to the obtaining of any necessary prior consents from the responsible authority. The Contractor shall comply with all requirements and restrictions that may be imposed and shall cover any costs involved. No instruction issued to the Contractor by the Architect shall relieve the Contractor from his responsibility for compliance with the Act. The Contractor shall provide for taking all reasonable precautions to ensure the efficient protection of all streams and waterways against pollution arising out of or by reason of the execution of the works. The Contractor shall provide for taking all necessary precautions to prevent nuisance from water, smoke, dust, rubbish and other causes.

3.11 *Maintain adjoining building.* The Contractor shall allow for taking all necessary precautions to prevent any damage occurring to the stone cottage abutting the eastern boundary of the site and to reinstate any damaged parts of the building arising from the works at his own expense.

3.12 *Maintain public and private roads.* The Contractor shall maintain public and private roads, footpaths, kerbs and channels, and shall keep the approaches to the site clear of mud and dirt. The Contractor shall make good at his own expense any damage caused by his own or any sub-contractor's or supplier's transport and shall pay any necessary costs and charges.

3.13 *Maintain live services.* The Contractor shall establish the positions of, protect, uphold and maintain all pipes, ducts, sewers, service mains, overhead cables and the like, during the execution of the works. The Contractor shall make good any damage due to any cause within his control at his expense or pay all resultant costs and charges. Where it is necessary to interrupt any such mains or services for the purpose of making temporary or permanent connections or disconnections, prior written permission shall be obtained from the Architect and, where appropriate, from the Local Authority or statutory undertaking, and the duration of any interruption kept to a minimum.

3.14 *Security.* The Contractor shall provide for carrying out adequate security arrangements designed to prevent vandalism and theft and generally to protect the works, both permanent and temporary, and all materials and goods on the site.

3.15 *Protection of work in all sections.* The Contractor shall provide for carefully covering up and protecting all sections of the works or any adjoining property exposed by these works from adverse weather or other factors.

3.16 *Restriction on use of radios.* The Contractor shall ensure that any radios of employees are used only when provided with ear plugs or head phones.

*Specific Limitations on Method/Sequence/Timing* (as SMM A35.1.1–8.1–2.0)

3.17 *Method and sequence of work.* The Contractor, upon acceptance of his offer, shall proceed immediately with the preparation of a programme or statement clearly identifying the sequence of all operations and the time limit within which the Contractor proposes that each operation shall be commenced and completed. The Contractor, in preparing his programme, shall be held to have co-ordinated the whole of the works embraced in this contract, including the work of nominated sub-contractors, local authorities and statutory undertakings, whether engaged by him or directly by the Employer. The Architect shall supply the Contractor with adequate details of all work to be carried out by others engaged directly by the Employer where such work is to be integrated with the work under this contract. Upon the agreement or negotiated amendment of the programme by the Architect, the Contractor shall be responsible for executing the works in conformity with it.

Two copies of the agreed programme are to be supplied to the Architect within seven days of agreement being reached. At the same time the Contractor shall issue to the Architect a statement setting out the latest dates by which drawings and other instructions are required by him for the implementation of the agreed programme. (Note: effective programming and progressing of the works and the supply of Architect's drawings at the proper times are vitally important to restrict the submission of Contractor's claims).

3.18 *Access.* Access to the site is restricted to the Black Street frontage.

3.19 *Use of the site.* The whole of the site shown on drawing HC 102/1 is available to the Contractor. No operatives shall trespass beyond the limits of the working area, other than to carry out works in connection with the contract. All temporary buildings, plant, spoil heaps and materials are to be sited in positions approved by the Architect. (Note: if the site is extremely confined it may be necessary to provide space for storage, huts and the like outside the boundaries of the site.)

3.20 *Use or disposal of materials found.* The Contractor shall not excavate on the site for sand or gravel, but any obtained from the excavations shall remain the property of the Employer and may, if approved by the Architect, be used in the work. The quantity so used shall be measured and the value assessed and deducted from the Contract Sum.

3.21 *Working hours.* Working hours are limited to the normal working hours defined in the National Working Rule 2 for the Building Industry. Overtime shall not be worked without the prior express permission in writing from the Architect.

*Facilities/Temporary work/Services* (as SMM A36.1.1–9.1–2.0

3.22 *Offices and sanitary accommodation.* The Contractor shall provide and maintain a secure, weatherproof office for the Clerk of Works with a floor area of not less than 20 m$^2$ and not less than 2.10 m high, with adequately insulated floor, walls and roof, the walls lined with fire resisting material and the floor covered with linoleum. He shall provide a desk or table with lockable drawers, clothes locker, ample plan chests or racks and sloping plan bench, chairs, lockable steel filing cabinet, drawing board, tee-square and lockable equipment cupboard. He shall also provide artificial lighting, adequate heating, attendance and cleaning.

The Contractor is to provide and maintain separate temporary toilet accommodation for the use of the Clerk of Works and other Employer's representatives and to maintain it in a clean and odour-free condition. Wherever possible such accommodation shall be of the flushing type and temporarily connected to a suitable drain. The Contractor shall provide a washbasin with an adequate supply of soap and clean towels.

The Contractor shall alter, move and adapt the foregoing temporary buildings as necessary and finally clear away and make good all surfaces and services disturbed. (Note: these requirements are based broadly on the recommendations of the Institute of Clerk of Works of Great Britain Incorporated on Clerks of Works Site Offices.)

3.23 *Temporary fences, hoardings, screens and roofs.* The Contractor shall provide temporary fences, hoardings, screens, roofs and the like for protecting the public, for the proper execution of the works and for meeting all statutory requirements. Where drain or service trenches are to be excavated outside the working area, such trenches shall be fenced or enclosed at all times.

3.24 *Name boards.* The Contractor shall, immediately, upon the commencement of the works and in accordance with details supplied by the Architect, make and erect a sign board showing the name and crest of the Employer and such other information as shall be directed by the Architect. The cost of the sign board including erection, taking down, removing and making good all work disturbed is covered by a provisional sum included elsewhere in the bill of quantities.

The Contractor may erect his name board in a position approved by the Architect, bearing his name and address and those of his sub-contractors. This board shall be supplied and erected at the Contractor's expense. No other notices or advertisements shall be permitted upon any part of the site.

3.25 *Technical and surveying equipment.* The Contractor shall provide all necessary surveying instruments and ancillary items required for setting out.

3.26 *Temperature and humidity.* The Contractor shall maintain a temperature of 21°C and a relative humidity of 55 per cent ambient. (Note: the Code of Procedure emphasises that rule A36.1.6 relates only to those cases where special requirements for temperature and humidity levels are imposed by the Employer, such as in computer rooms or telephone exchanges or where there are special finishings. The attainment and maintenance of suitable levels for normal construction are the responsibility of the Contractor.)

3.27 *Telephone.* The Contractor is responsible for the installation of a separate telephone for the use of the Employer and for the cost of telephone calls made on his behalf. (Note: a provisional sum for installation and payment of calls is included elsewhere in the bill of quantities).

3.28 *Winter building.* The Contractor shall be familiar with the measures described in *Winter Building*, DOE (HMSO) for maintaining continuity of working and productivity during adverse weather conditions. The Contractor shall provide for taking the measures described as and when desirable and practicable, including the provision of site lighting, shelters and screens for the protection of operatives and the works and the protection of materials and plant from the effects of the weather.

*Operation/Maintenance of the Finished Building* (as SMM A37.1.1.1–2.0)

3.29 *Operation of services.* The Contractor shall take responsibility for operating the hot water and heating systems, electrical installation, fire alarms and emergency lighting throughout the defects liability period.

## CONTRACTOR'S GENERAL COSTS

SMM A40 to A44 list typical contractor's general cost items for ease of pricing by the estimator as fixed and/or time related charges. They cover a wide range of items as shown in the following list.

(1) Management and staff which include management personnel, trades supervision, engineering, programming and production staff, quantity surveying support staff and the like (SMM A40).

(2) Site accommodation which includes offices, laboratories, cabins, stores, compounds, canteens, sanitary facilities and the like (SMM A41).

(3) Services and facilities, including such items as power, lighting, water, telephone, safety, health and welfare, storage of materials, rubbish disposal, cleaning, drying out, protection of work in all sections, security, maintenance of public and private roads, small plant and tools, and general attendance on nominated sub-contractors (SMM A42.1.1–16).

General attendance is deemed to include the use of the Contractor's temporary roads, pavings and paths, standing scaffolding, standing power operated hoist plant, the provision of temporary lighting and water supplies, clearing away rubbish, provision of space for the sub-contractor's own offices and the storage of his plant and materials and the use of messrooms, sanitary accommodation and welfare facilities provided by the Contractor (SMM A42.C3). This definition includes the changes listed in the September 1988 amendment to SMM7.

(4) Mechanical plant which includes such items as cranes, hoists, personnel transport, transport, earthmoving plant, concrete plant, piling plant and paving and surfacing plant (SMM A43.1.1–9).

(5) Temporary works which include such items as temporary roads, temporary walkways, access scaffolding, support scaffolding and propping, hoardings, fans and fencing, hardstanding and traffic regulations (SMM A44.1.1–8).

## 4.00 Contractor's General Cost Items

*Management and Staff* (as SMM A40.1.0.1–2.0)

4.01 *Site administration.* The Contractor is to provide for all site administration costs including the cost of the person-in-charge in accordance with clause 10 of the Conditions of Contract.

*Site Accommodation* (as SMM A41.1.0.1–2.1)

4.02 *Site accommodation.* The Contractor shall provide and maintain, in positions to be agreed with the Architect, temporary office accommodation for the Contractor's staff and watertight sheds for the storage of materials, tools and tackle and for the use of operatives. (Note: sheds and the like for the use of nominated sub-contractors are given in the items of special attendance following the relevant prime cost items. Similar items follow for toilets and final clearance as in item 3.22.)

*Services and Facilities* (as SMM A42.1.1–16.1–2.1)

4.03 *Power and lighting.* The Contractor shall provide all artificial lighting and power for use on the works, pay all necessary charges, provide all temporary connections, leads and fittings and shall clear away and make good on completion. (Note: the provision of power for the use of nominated sub-contractors, where required, is included in the items of special attendance on nominated sub-contractors following the relevant prime cost items.)

4.04 *Water.* The Contractor shall provide clean, fresh water for use on the works, pay all necessary charges, provide all temporary storage, plumbing services and connections and shall clear away and make good on completion (carried to General Summary).

4.05 *Telephones and administration.* The Contractor shall provide and maintain a telephone service to the office of the person-in-charge and to the offices of such other of the Contractor's site staff as may be necessary for the full period of the works and shall pay all appropriate charges and expenses.

4.06 *Safety, health and welfare.* The Contractor shall provide for all costs incurred by and comply with all safety, health and welfare regulations appertaining to all operatives on the site, including the provision of protective clothing. Any accommodation required shall be cleared away on completion and the area made good.

4.07 *Rubbish disposal.* The Contractor shall provide for the removal of all rubbish from the site both as it accumulates from time to time and on completion.

4.08 *Cleaning.* The Contractor shall provide for keeping the site tidy throughout the construction period, for cleaning the buildings inside and out, touching up paintwork and for leaving the whole of the works clean on completion to the satisfaction of the Architect.

4.09 *Drying out.* The Contractor shall provide for adequate drying out of the buildings during the execution of the works both for the Contractor's own work and for the work of all nominated sub-contractors and shall allow for providing all necessary labour, appliances and fuel.

4.10 *Protection of work in all sections.* The Contractor shall provide for carefully covering up and protecting all sections of the works or any adjoining property exposed by these works from inclement weather or other harmful sources. (See also item 3.15.)

4.11 *Security.* The Contractor shall safeguard the works, materials and plant against damage or theft, including the provision of all necessary watching and lighting. (See also item 3.14.)

4.12 *Maintain public and private roads.* The Contractor shall maintain public and private roads, footpaths, kerbs and channels, and shall keep the approaches to the site clear of mud and dirt. The Contractor shall make good at his own expense any damage caused by his own or any sub-contractor's or supplier's transport and shall pay any necessary costs and charges. (See also item 3.12.)

4.13 *Small plant and tools.* The Contractor shall provide for all necessary small plant and tools required in the execution of the works.

4.14 *Attendance on site.* The Contractor shall provide for the provision of a chainman and any special attendance required by the Clerk of Works.

4.15 *Rectification of minor defects.* The Contractor shall allow for the cost of labour and materials in rectifying minor defects prior to handover of the works and during the defects liability period.

4.16 *Disbursements arising from the employment of workpeople (carried to General Summary).* The Contractor shall provide for all costs in respect of workpeople for:
(a) National insurance contributions
(b) Pensions
(c) Annual and public holidays
(d) Travelling time, expenses, fares and subsistence (see item 4.19 for personnel transport)
(e) Guaranteed time
(f) Non-productive time and other expenses in connection with overtime
(g) Any incentive and bonus payments
(h) Severance pay and obligations under the Redundancy Payments Acts
(i) Construction Industry Training Board levies
(j) Disbursements under the Sick Payments Scheme
(k) Any other disbursements arising from the employment of workpeople (see General Summary)

4.17 *General attendance on nominated sub-contractors.* The Contractor shall allow for the cost of providing general attendance on all nominated sub-contractors. (Note: the Code of Procedure recommends that a single item shall be provided in the preliminaries bill for general attendance on all nominated sub-contractors.)

*Mechanical Plant* (as SMM A43.1.1–9.1–2.1)

4.18 *Hoists.* The Contractor shall allow for the costs of the provision and use of all necessary hoists for lifting materials and components on the site.

4.19 *Personnel transport.* The Contractor shall provide for all costs in respect of transport of personnel to and from the site.

4.20 *Transport.* The Contractor shall provide for the costs of all necessary transport required for the conveyance of materials, components and equipment to be used in the execution of the works.

4.21 *Earthmoving plant.* The Contractor shall allow for the costs of the provision and use of all necessary earthmoving plant to carry out the excavating and filling operations on the site.

4.22 *Concrete plant.* The Contractor shall allow for the costs of the provision and use of all concrete plant required for the execution of the works.

4.23 *Paving and surfacing plant.* The Contractor shall allow for the costs of the provision and use of the necessary paving and surfacing plant required for the construction of the building and external works.

*Temporary Works* (as SMM A44.1.1–8.1–2.1)

4.24 *Temporary roads.* The Contractor shall provide for and maintain all necessary temporary roads and shall clear away and make good on completion. (Note: temporary roads required for the use of nominated sub-contractors are given in the items of special attendance following the relevant prime cost items.)

4.25 *Access scaffolding.* The Contractor shall provide all necessary access scaffolding for the execution of the works. (Note: Scaffolding required additional to the Contractor's standing scaffolding, or standing scaffolding required to be altered or retained, is given in the items of special attendance following the relevant prime cost items as SMM A51.D8).

4.26 *Hoardings, fans, fencing, etc.* The Contractor shall provide and maintain all necessary temporary hoardings, fans, fencing and the like for the proper execution of the works, for the protection of the public and the occupants of adjoining premises and for meeting the requirements of any local or other authority, and shall alter, remove, adapt as necessary and finally clear away when no longer required and shall reinstate all surfaces and services disturbed. The Contractor is referred to item 3.23 for particulars of temporary fences and hoardings that he is required by the Employer to provide.

4.27 *Hardstanding.* The Contractor shall provide and maintain all necessary hardstandings, and shall clear away and make good on completion. (Note: hardstandings required for the use of nominated sub-contractors are given in the items of special attendance following the relevant prime cost sums.)

4.28 *Traffic regulations.* The Contractor shall comply with all traffic regulations, particularly those relating to the loading and unloading of vehicles and shall provide any necessary traffic control.

## WORKS BY NOMINATED SUB-CONTRACTORS, GOODS AND MATERIALS FROM NOMINATED SUPPLIERS AND WORKS BY STATUTORY AUTHORITIES

*Works by Nominated Sub-contractors*

Works to be carried out by nominated sub-contractors are to be given as prime cost sums, describing the work and making provision for the insertion of a percentage addition for main contractor's profit (SMM A51.1.1–2.1.0). General attendance on nominated sub-contractors, as listed in SMM A42.1.16, is given elsewhere as a separate single item. The list provided in SMM A42.C3 is intended to give an indication of the kind of facilities that are normally made available to sub-contractors where they are provided by the Contractor to meet his own needs. Further items are required for special attendance and details of these are listed in SMM A51.1.3.1–8.1–2. This procedure enables proper provision to be made for significant costs beyond those envisaged in the definition of general attendance, and details of these items must be stated and they can be priced by the Contractor using fixed and/or time related charges. An example follows later in the chapter.

The Code of Procedure recommends that bills of quantities used for inviting tenders from potential nominated sub-contractors should be drawn up in accordance with SMM7 as a whole, as if the work was main contractor's work, and should include preliminaries and relevant drawings. As much information as possible should be given in respect of nominated sub-contractors' work in order that tenderers can make proper allowance when assessing the overall programme and establishing the contract period if not already prescribed. A list describing the component elements of the work and the extent and possible value of each element would help the sub-contractor with pricing. The location of the main plant, for example, whether in the basement or on the roof will influence tenderers' programmes. The Code points out that it is good practice to seek programme information when obtaining estimates from sub-contractors so that this can be incorporated in the bills of quantities, for the benefit of tenderers.

The rule for *special attendance* in SMM A51.1.3 is included to enable proper provision to be made for costs beyond those envisaged in the definition of general attendance. Special scaffolding or scaffolding additional to the main contractor's standing scaffolding required for use by sub-contractors should be described. For example, windows supplied and fixed by a nominated sub-contractor who requires scaffolding for fixing will require a bill item giving the dimensions for each elevation. It is insufficient merely to refer to the items listed in the third column (SMM A51.1.3.1–8), as details of requirements should be stated. Where adequate information cannot be provided a provisional sum should be used.

*Materials and Goods from Nominated Suppliers*

Materials and goods that are to be obtained from a nominated supplier are given as a prime cost sum, giving a suitable description and provision for the insertion of a percentage addition for the main contractor's profit, with the fixing dealt with in accordance with the normal rules in the Standard Method (SMM A52.1.1–2.1.0). An example follows later in the chapter.

*Works by Statutory Authorities*

Provisional sums are to be included for work to be carried out by a local authority or a statutory undertaking such as the construction of a permanent footpath crossing to the site by a local authority (SMM A53.1.1.0.0), or the laying of a new mains gas supply to the site (SMM A53.1.2.0.0).

Provisional sums can be inserted for defined or undefined work. A provisional sum for defined work is a sum provided for work which is not completely designed but for which the following information shall be provided:

(a) The nature and construction of the work.
(b) A statement of how and where the work is to be fixed to the building and what other work is to be fixed thereto.
(c) A quantity or quantities which indicate the scope and extent of the work.
(d) Any specific limitations and the like identified in Section A35.

In these circumstances the Contractor will be deemed to have made due allowance in programming, planning and pricing preliminaries. In provisional sums for undefined work the above information cannot be supplied (SMM General Rules 10.3 to 10.5).

# PRIME COST AND PROVISIONAL SUMS AND DIRECT WORKS

The following examples illustrate a suitable format for each of these items, which are usually inserted after the work section bills, and the collected monetary totals carried from the bottom of each page to the General Summary.

| | BILL Nr 6 | | | | |
|---|---|---|---|---|---|
| | PC AND PROVISIONAL SUMS AND DIRECT WORKS PC SUMS – NOMINATED SUPPLIERS  Provide the following sums for materials and goods to be supplied and delivered to site: | | | | |
| A | Aluminium windows (Aluminium Window Co. Ltd) comprising 25 nr individual windows; total area 200 m². | Item | | 2550. | 00 |
| B | Profit. | | % | | |
| C | Carpeting (Axminster Carpets Ltd); 330 m². | Item | | 3960. | 00 |
| D | Profit. | | % | | |
| | PC SUMS — NOMINATED SUB-CONTRACTORS  Provide the following sum for Works to be carried out: | | | | |
| A | Electrical engineering installation comprising sub-mains and distribution board, lighting and power installations and electrical work associated with mechanical services installations. | | | 9900. | 00 |
| B | Profit. | Item | % | | |
| C | Providing hardstanding for sub-contractor's own offices and storage hut. | Item | | | |
| D | Providing power, maximum load 7.5 kVA. | Item | | | |
| | PROVISIONAL SUMS  Provide the following sums for work or costs that cannot be entirely foreseen, defined or detailed: | | | | |
| A | Fees and charges including rates on temporary buildings that the Contractor is required to pay to local authorities. | Item | | 750. | 00 |
| B | Special testing of materials that the Architect may direct the Contractor to have carried out in accordance with clause 8 of the Conditions of Contract. | Item | | 230. | 00 |
| C | Maintaining insurances to a limit of £750 000 indemnity in respect of any expense, liability, loss, claim or proceedings that the Employer may incur or sustain by reason of damage to any property other than the Works, all as described in clause 21.2 of the Conditions of Contract. | Item | | 600. | 00 |
| D | Maintaining the required temperature and humidity levels. | Item | | 900. | 00 |
| E | Installation of telephone for the use of the Employer and the cost of all calls made on his behalf. | Item | | 450. | 00 |
| F | Contingencies. | Item | | 3700. | 00 |
| | DIRECT WORKS  The following works will be carried out by others directly employed by the Employer  (Note: Items of attendance are to be priced and extended into the money columns. The sums representing the Value of the Direct Works are NOT to be similarly extended NOR included in the Tender.) | | | | |
| A | Fixed furniture by Furniture Supplies Ltd (thirteen thousand two hundred pounds). | | | | |
| B | Providing temporary power and lighting. | Item | | | |
| C | Clearing away rubbish. | Item | | | |

(Note: the alphabetical reference letters normally restart at A at the top of each page to assist in identifying items by letter and page number.)

# DAYWORKS

Dayworks can be included in the same bill with provisional sums and prime cost items and it may also include builder's work in connection with works by nominated sub-contractors. Alternatively, Dayworks may form a separate bill. A specimen Dayworks bill follows using the approach adopted by the Society of Chief Quantity Surveyors in Local Government.

| | BILL Nr 7 | | | | |
|---|---|---|---|---|---|
| | DAYWORKS | | | | |
| | Provide the following sums for works or costs that cannot be entirely foreseen, defined or detailed and for which a valuation shall be made in accordance with clause 13.5.4.1 of the Conditions of Contract | | | | |
| A | Prime cost of labour as defined (eight hundred pounds). | Item | | 800. | 00 |
| B | Percentage addition for overheads as defined and profit. | | % | | |
| C | Prime cost of materials and goods as defined (three hundred pounds). | Item | | 300. | 00 |

| | | | | | | |
|---|---|---|---|---|---|---|
| | D | Percentage addition for overheads as defined and profit. | | % | | |
| | E | Prime cost of plant in accordance with the Schedule of Basic Plant Charges published by the Royal Institution of Chartered Surveyors, January 1981 (one hundred and fifty pounds). | Item | | 150. | 00 |
| | F | Percentage addition for overheads as defined and profit. | | % | | |
| | | Carried to GENERAL SUMMARY | | £ | | |

## GENERAL SUMMARY

The General Summary comprises Preliminaries, the totals carried from each of the work sections, prime cost sums, provisional sums and direct works, dayworks and the four items indicated in the Preliminaries, namely 2.03 (clause 21), 2.03 (clause 22A), 4.04 and 4.16 relating to insurances, water and disbursements arising from the employment of workpeople. The total sum in the General Summary constitutes the Amount of the Tender.

## PREAMBLES

Preambles are clauses inserted at the head of each work sectional bill covering matters relating to the specific work section that should be brought to the attention of the estimator as they will affect the rates that he inserts against billed items of measured work. They almost invariably incorporate material descriptions and particulars of the standard or quality of workmanship to be employed.

The use of preambles enables the descriptions of the measured work items to be reduced significantly in length and assists considerably with their pricing. In practice the preamble clauses are printed or typed across the description, quantity and unit columns of the bill pages. Another alternative is to keep all the preamble clauses in all work sections together in a separate bill following the Preliminaries Bill. Although this is becoming a quite common practice it is rather less convenient for the Contractor.

The headings of the principal preamble clauses in each work section are listed, followed by specimen preamble clauses in selected work sections, in order that the student is made familiar with their normal format and content. It will be noted that preamble clauses are often subdivided into three distinct sections — general; materials; and workmanship. In the preamble clause headings that follow the three categories of item have been identified by the use of prefixes — A for general items, B for materials and C for workmanship, in addition to the SMM work section letter in the clause references. In preparing the preambles section, reference has been made to *Preambles to Bills of Quantities*, GLC, Architectural Press, adjusted as necessary to conform to *SMM7* and the *Common Arrangement*.

### TYPICAL PREAMBLE CLAUSE HEADINGS

*Demolition/Alteration/Renovation (SMM Section C)*

| | |
|---|---|
| CA1 | General particulars |
| CB2 | Old materials to be retained by the Employer |
| CB3 | Old materials to be re-used |
| CC4 | Temporary screens and the like |
| CC5 | Shoring and support |
| CC6 | Protection of existing work, fittings and adjoining property |
| CC7 | Disconnection of electricity, gas and water supplies prior to demolition work being commenced |

*Groundwork (SMM Section D)*

*Excavating and Filling (D20)*

| | |
|---|---|
| DA1 | Site preparation |
| DA2 | Datum, site and floor levels |
| DA3 | Trial holes and ground conditions |
| DA4 | Groundwater level |
| DA5 | Unauthorised excavations |
| DA6 | Re-use of excavated material on site |
| DB7 | Hardcore and blinding |
| DB8 | Topsoil |
| DB9 | Grass seed |
| DC10 | Bottoms of excavations to be approved |
| DC11 | Disposal of excavated material |
| DC12 | Excavation below required levels |
| DC13 | Earthwork support |
| DC14 | Filling |
| DC15 | Preparation of cultivated areas |
| DC16 | Fertilising and grass seeding |

*Piling (D30–31)*

| | |
|---|---|
| DB17 | Type of piling |
| DC18 | Level at which piling is to start |
| DC19 | Precautions to be taken to adjoining properties |
| DC20 | Builder's work (where piling carried out by a specialist) |

*In situ Concrete (SMM Section E)*

| | |
|---|---|
| EA1 | Bar bending schedules |
| EB2 | Cement |
| EB3 | Aggregates |
| EB4 | Reinforcement |
| EB5 | Expansion joint filler |
| EB6 | Waterproof building paper |
| EB7 | Precast concrete components |
| EB8 | Water |
| EC9 | Storage of materials |
| EC10 | Nominal and designed mixes |

*Preliminaries, Preambles, Co-ordinated Project Information and Use of Computers in Bill Production*     13

| | |
|---|---|
| EC11 | Testing of materials |
| EC12 | Mixing of concrete |
| EC13 | Formwork |
| EC14 | Placing and compaction of concrete |
| EC15 | Construction joints |
| EC16 | Curing and striking times |
| EC17 | Concreting in cold weather |
| EC18 | Concrete roads and pavings |

## Masonry (SMM Section F)

| | |
|---|---|
| FA1 | Samples |
| FA2 | Testing |
| FB3 | Bricks and blocks generally |
| FB4 | Common bricks |
| FB5 | Facing bricks |
| FB6 | Fixing bricks |
| FB7 | Flue linings and terminals |
| FB8 | Precast concrete blocks |
| FB9 | Hollow clay blocks |
| FB10 | Stone and/or quarry |
| FB11 | Air bricks |
| FB12 | Portland cements |
| FB13 | Sulphate resisting Portland cement |
| FB14 | Masonry cement |
| FB15 | Sand for mortar |
| FB16 | Water |
| FB17 | Bitumen for damp-proof courses |
| FB18 | Wall ties |
| FB19 | Expansion joint material |
| FB20 | Cramps |
| FB21 | Dowels |
| FB22 | Joggles |
| FC23 | Storage of materials |
| FC24 | Faced work to be kept clean |
| FC25 | Work in cold weather |
| FC26 | Wetting bricks |
| FC27 | Mortar batching and mixing |
| FC28 | Table of mortar mixes |
| FC29 | Hollow walls |
| FC30 | Damp-proof courses |
| FC31 | Jointing and pointing |
| FC32 | Texture and finish of stone |
| FC33 | Bonding of stone to backing |
| FC34 | Treatment of back of stone |

## Structural/Carcassing Metal/Timber (SMM Section G)

### Structural steel framing (G10)

| | |
|---|---|
| GA1 | Standard of construction |
| GA2 | Fabrication by specialist firm |
| GA3 | Shop details |
| GA4 | Erection arrangements |
| GB5 | Quality of steel |
| GB6 | Marking of steel |
| GB7 | Standard dimensions and weight of steel |
| CB8 | Condition of surfaces |
| GB9 | Tests and inspection |
| GB10 | Metallic coatings |
| GB11 | Paint |
| GC12 | Workmanship generally |
| GC13 | Fabrication |
| GC14 | Joints and connections |
| GC15 | Painting at works |

### Carpentry/timber framing/first fixing (G20)
(Note: T inserted in prefix for timber)

| | |
|---|---|
| GTA1 | Definitions and timber generally |
| GTA2 | Fixing accessories |
| GTA3 | Storage |
| GTB4 | Softwood for carpentry |
| GTB5 | Structural plywood |
| GTB6 | Metalwork |
| GTB7 | Trussed rafters |
| GTC8 | Screwed or bolted joints and connections |
| GTC9 | Longitudinal joints |
| GTC10 | Trimming |
| GTC11 | Pitched roofs: nailing |
| GTC12 | Pitched roofs: erection of trussed rafters |
| GTC13 | Fixing joist hangers |
| GTC14 | Firrings for flat roofs |
| GTC15 | Preservation of structural timber |

### Edge supported/reinforced woodwool slab decking (G32)
(Note: D inserted in prefix for decking)

| | |
|---|---|
| GDB1 | Woodwool slabs for decking |
| GDB2 | Woodwool kerbs |
| GDB3 | Woodwool angle fillets |
| GDC4 | Storing and protecting woodwool slabs |
| GDC5 | Fixing woodwool slabs |

## Cladding/Covering (SMM Section H)

### Clay/concrete roof tiling (H60)

| | |
|---|---|
| HA1 | Samples |
| HB2 | Plain tiles and fittings |
| HB3 | Single lap tiles and fittings |
| HB4 | Ridge terminal units |
| HB5 | Nails |
| HB6 | Battens |
| HB7 | Underlay |
| HB8 | Hip irons |
| HC9 | Fixing underlay |
| HC10 | Fixing battens |
| HC11 | Tiling |
| HC12 | Eaves |
| HC13 | Verges |
| HC14 | Ridges and hips |

### Metal sheet flashings and gutters (H71–74)
(Note: M inserted in prefix for metal)

| | |
|---|---|
| HMB1 | Lead |
| HMB2 | Aluminium |
| HMB3 | Copper |
| HMB4 | Zinc alloy |
| HMB5 | Mastic jointing to prefabricated aluminium flashings |
| HMC6 | Expansion jointing to prefabricated aluminium flashings |

*Waterproofing (SMM Section J)*

*Mastic asphalt tanking/damp-proof membranes and mastic asphalt roofing/insulation/finishes (J20 and J21)*

| | |
|---|---|
| JA1 | Surfaces and screeds |
| JA2 | Guarantee |
| JA3 | Testing |
| JB4 | Bitumen coated lightweight screed |
| JB5 | Vapour barrier |
| JB6 | Asphalt for roofing |
| JB7 | Asphalt for tanking |
| JB8 | Felt underlay |
| JB9 | Reinforcement |
| JB10 | Dressing compound for chippings |
| JB11 | Chippings |
| JC12 | Laying felt vapour barrier |
| JC13 | Laying bitumen coated lightweight screed |
| JC14 | Laying asphalt |
| JC15 | Joints, fillets, skirtings and upstands |
| JC16 | Felt underlay |
| JC17 | Surfacing roofs with chippings |
| JC18 | Reinforcement |
| JC19 | Tests for falls |
| JC20 | Temporary protection |

*Built up felt roof coverings (J41)*
(Note: B inserted in prefix for bitumen)

| | |
|---|---|
| JBA1 | Statement as to screed and underbed |
| JBA2 | Guarantee |
| JBA3 | Samples |
| JBB4 | Primer |
| JBB5 | Bonding compound |
| JBB6 | Dressing compound for chippings |
| JBB7 | Chippings |
| JBB8 | Felt for mineral surfaced finish |
| JBB9 | Felt for chipping finish |
| JBB10 | Felt for gutters |
| JBB11 | Nails |
| JBC12 | Preparation |
| JBC13 | Falls |
| JBC14 | Laps and bonding |
| JBC15 | Surfacing roofs with chippings |

*Linings/Sheathing/Dry Partitioning (SMM Section K)*

*Timber board flooring/sheathing/linings/casings (K20)*

| | |
|---|---|
| KB1 | Tongued and grooved softwood boarding |
| KB2 | Wrought softwood boarding |
| KB3 | Hardwood strip boarding |
| KB4 | Plywood |
| KB5 | Chipboard |
| KB6 | Fibre building boards |
| KB7 | Blockboard and laminboard |

*Note:* these clauses could be followed by dry linings and demountable partitions.

*Windows/Doors/Stairs (SMM Section L)*

| | |
|---|---|
| LB1 | Joinery timber generally |
| LB2 | Softwood for joinery |
| LB3 | Hardwood |
| LB4 | Adhesives |
| LB5 | Flush doors |
| LB6 | Firecheck flush doors |
| LB7 | Panelled and glazed doors |
| LB8 | Door frames and linings |
| LB9 | Wood windows |
| LB10 | Sub-frames to metal windows |
| LB11 | Aluminium windows |
| LB12 | Steel windows |
| LB13 | Staircases |
| LC14 | Preservation of external joinery |
| LC15 | Workmanship in joinery generally |
| LC16 | Arrises |
| LC17 | Priming |

*General glazing (L40)* (Note: G inserted in prefix for glazing)

| | |
|---|---|
| LGA1 | Samples |
| LGB2 | Glass |
| LGB3 | Laminated safety glass |
| LGB4 | Factory made double glazing units |
| LGB5 | Extruded aluminium glazing cap |
| LGB6 | Linseed oil putty for glazing to wood |
| LGB7 | Metal casement putty |
| LGB8 | Non-setting compounds for glazing with beads |
| LGC9 | Storage and care of glass |
| LGC10 | Rebates and beads |
| LGC11 | Use of glazing compounds |
| LGC12 | Putty glazing to wood or metal frames |
| LGC13 | Bead glazing |

*Surface Finishes (SMM Section M)*

*Sand cement/concrete/granolithic screeds/flooring (M10)*

| | |
|---|---|
| MB1 | Cement |
| MB2 | Sand |
| MB3 | Water |
| MC4 | Storage, testing, frosty weather and mixing of materials |
| MC5 | Cement and sand proportions |
| MC6 | Preparation of surfaces |
| MC7 | Bay sizes and movement joints |
| MC8 | Laying screeds |
| MC9 | Surface of screeds |
| MC10 | Curing |

*Note:* these clauses could be followed by other *in situ* finishings but terrazzo work would be taken later (M41).

*Plastered/rendered/roughcast coatings (M20)*
(Note: R inserted in prefix for rendering)

| | |
|---|---|
| MRB1 | Metal lathing |
| MRB2 | Gypsum plaster lath |
| MRB3 | Cements |
| MRB4 | Gypsum plasters |
| MRB5 | Bonding plaster |
| MRB6 | Lightweight plasters |
| MRB7 | Sands |
| MRB8 | Water |
| MRB9 | Angle and casing beads |
| MRC10 | Storage of materials |
| MRC11 | Testing |
| MRC12 | Preparation of surfaces |
| MRC13 | Spatterdash |

| | |
|---|---|
| MRC14 | Dubbing out |
| MRC15 | Fixing metal lathing |
| MRC16 | Fixing plaster lath and filling joints |
| MRC17 | Frosty weather |
| MRC18 | Mixing of materials |
| MRC19 | Period between coats |
| MRC20 | Finish |
| MRC21 | Junction of wall and ceiling plaster |
| MRC22 | Arrises |
| MRC23 | Tyrolean finish |
| MRC24 | Table of plastering mixes |

*Note:* These clauses could be followed by the requirements for any fibrous plaster.

*Tile, slab and block finishes to walls and floors (M40/41/42/50)* (Note: T inserted in prefix for tiling)

| | |
|---|---|
| MTA1 | Samples |
| MTB2 | Glazed ceramic wall tiles |
| MTB3 | Quarry tiles |
| MTB4 | Concrete tiles |
| MTB5 | Precast terrazzo |
| MTB6 | Vinylised thermoplastic/vinyl/vinyl asbestos tiles |
| MTB7 | Cork tiles |
| MTB8 | Wood blocks |
| MTB9 | Mortar for bedding and pointing |
| MTC10 | Preparation |
| MTC11 | Laying finishings |

*Note:* These clauses could be followed by the requirements for any flexible sheet finishings and edge fixed carpeting

*Painting/clear finishing (M60)*
(Note: P inserted in prefix for painting)

| | |
|---|---|
| MPA1 | Approval of brands |
| MPA2 | Colour range |
| MPA3 | Storage |
| MPA4 | Testing and use of materials |
| MPB5 | Knotting |
| MPB6 | Stopping and fillers |
| MPB7 | Masonry paint |
| MPB8 | Emulsion paint |
| MPB9 | Black bituminous paint |
| MPB10 | Chlorinated rubber paint |
| MPB11 | Primers for various surfaces, such as alkaline, aluminium, bituminous, iron and steel, zinc and galvanised steel, hardboard, fibreboard, plasterboard, softwood and hardwood. |
| MPB12 | Oil paints |
| MPB13 | Heat resisting paint |
| MPB14 | Anti-condensation paint |
| MPB15 | Alkyd varnish |
| MPB16 | Polyurethane varnish |
| MPB17 | Exterior wood stains |
| MPB18 | Wallpaper and wall coverings |
| MPC19 | Preparation of surfaces, such as those covered by MPB11 |
| MPC20 | Adherence to manufacturer's instructions |
| MPC21 | Brushwork |
| MPC22 | Priming of joinery |
| MPC23 | Rubbing down |
| MPC24 | Differing colours of undercoats |
| MPC25 | Painting in unsuitable conditions and on wet surfaces |
| MPC26 | Damage to adjoining surfaces |
| MPC27 | Removal of ironmongery |

*Building Fabric Sundries (SMM Section P)*

*Insulation (P10/11)*

| | |
|---|---|
| PB1 | Insulating loose wool |
| PB2 | Insulating slab for thermal insulation (timber flat roof spaces) |
| PB3 | Expanded polystyrene board for thermal insulation |
| PB4 | Insulating quilt for sound insulation (solid floors) |
| PB5 | Expanded polystyrene board for sound insulation (solid floors) |
| PB6 | Foamed cavity wall insulation |
| PB7 | Bead cavity wall insulation |

*Ironmongery (P21)*

| | |
|---|---|
| PB8 | Ironmongery |

*Paving/Planting/Fencing/Site Furniture (SMM Section Q)*

*Fencing (Q40)*

| | |
|---|---|
| QB1 | Open type fencing |
| QB2 | Close type fencing |
| QB3 | Welded galvanised steel fencing |
| QB4 | Gates |
| QC5 | Erection of fencing |

*Disposal Systems (SMM Section R)*

*Rainwater pipework/gutters and foul drainage above ground (R10/11)*

| | |
|---|---|
| RB1 | Cast iron gutters and fittings |
| RB2 | Plastics gutters and fittings |
| RB3 | Cast iron pipes and fittings |
| RB4 | Plastics pipes and fittings |
| RB5 | Balloon guards |
| RB6 | Joints generally |
| RB7 | Testing |

*Drainage below ground/land drainage (F12/13)*
(Note: D is inserted in prefix for drainage)

| | |
|---|---|
| RDA1 | Notices, byelaws and regulations |
| RDA2 | Levels of existing drains |
| RDA3 | Surface loads |
| RDB4 | Granular material for bedding and surrounds |
| RDB5 | *In situ* concrete |
| RDB6 | Clay pipes and fittings |
| RDB7 | Precast concrete pipes and fittings |
| RDB8 | Cast iron spigot and socket drainpipes and fittings |
| RDB9 | Pitch fibre pipes and fittings |
| RDB10 | Land drainpipes |
| RDB11 | Gullies |
| RDB12 | Precast concrete manholes and inspection chambers |
| RDB13 | Bricks |
| RDB14 | Manhole and inspection chamber covers |
| RDB15 | Manhole step irons |
| RDC16 | Setting out |
| RDC17 | Excavation |
| RDC18 | Laying and jointing pipes |
| RDC19 | Pipe surrounds |
| RDC20 | Filling excavations |
| RDC21 | Construction of manholes |
| RDC22 | Gullies and gratings |

RDC23   Sealing and marking ends of drains for future connections
RDC24   Testing
RDC25   Protection

*Mechanical and Electrical Services (SMM Section Y)*

*Piped supply systems: cold and hot water (Y10/11/20/50)*

YA1    Acts, byelaws and notices
YB2    Copper tubes and fittings
YB3    Plastic overflow pipes
YB4    Chromium plating on copper tubes and fittings
YB5    Sleeves
YB6    Brasswork
YB7    Tanks and cylinders
YC8    Pipework generally
YC9    Thermal insulation
YC10   Testing

*Note:* mechanical and electrical services have been omitted as the information would normally be provided by specialists. The specification notes incorporated in chapters 8 and 9 provide useful guidelines to these work categories.

## SPECIMEN PREAMBLE CLAUSES

The following preamble clauses are intended to show the form and scope of a suitable range of preamble clauses. Limitations of space restrict the number that can be included and the two work sections of 'Excavating and Filling' and 'Structural Steel Framing' have been selected as giving the best spread and probably having the greatest applicability to the subjects covered in the final examinations set by the professional bodies and higher educational establishments.

The author has drawn extensively from the very comprehensive GLC, Architectural Press publication *Preambles to Bills of Quantities* in drafting the preamble clauses that follow. The GLC clauses have however been adapted or condensed where considered appropriate to make them more applicable to a specific project outside London. The reader is referred to the GLC publication for the full wording of the recommended preamble clauses for these and other work sections.

## EXCAVATING AND FILLING
*(SMM Work Section D20)*

### Generally

D1 *Site preparation*
All obstructions shall be reported to the Architect, following which they shall be removed or otherwise as directed by the Architect. All live drains, electric cables, gas and water services exposed during excavations shall be reported to the Architect.

The Contractor shall remove all trees and shrubs as indicated and agreed by the Architect, and they are to be removed without damage to adjoining trees, shrubs and buildings that are to remain. All trees and shrubs to be preserved shall be protected by adequate fencing until the completion of the permanent works, and extending sufficiently to protect the perimeter of branches and roots.

D2 *Datum, site and floor levels*
The Contractor shall establish and maintain on site a datum level working from the Ordnance bench mark on drawing HC 102/1. Immediately after receipt of the order to commence, the Contractor shall carry out and record a check level grid of the site which shall be agreed between the Architect and the Contractor within one week of the order being given. The Contractor shall work to the floor levels shown on drawing HC 102/2.

D3 *Trial holes and ground conditions*
Details of trial holes are shown on drawing HC 102/6. The ground generally consists of topsoil 150 mm thick and soft clay 2.00 m deep, overlying sand and gravel.

D4 *Groundwater level*
Groundwater level was established on 6 July 1989 as 4.60 m below existing ground level (124.80).

D5 *Unauthorised excavations*
The Contractor is prohibited from carrying out excavations other than those required for the execution of the Works and approved by the Architect. In the event of the Contractor making unauthorised excavations he will be required to backfill them with such materials as directed at his own expense.

D6 *Re-use of excavated material on site*
Excavated material complying with clause D14 may be used for filling to make up levels and for backfilling around foundations. Hard excavated material and brick rubble complying with clause D7 may be used for hardcore. Material for re-use must be kept in separate heaps free from contamination.

### Materials

D7 *Hardcore and blinding*
Hardcore shall be well graded hard brick, stone or gravel rejects to pass a 100 mm ring, free from all rubbish. The blinding for polythene membranes shall be clean soft sand.

D8 *Topsoil*
The topsoil for cultivated areas shall be good quality medium top spit loam, easily moulded when moist. It shall

be free from subsoil, pollutants, obnoxious weeds, roots, turf, couch grass, rubbish or an excessive proportion of clay, sand, gravel, chalk or lime. It shall not contain stones in quantities exceeding 10 per cent of the total bulk and those present shall not have any dimension exceeding 50 mm. Where topsoil is to be imported, the Contractor shall submit a sample load for approval and shall ensure that all subsequent deliveries are similar to the approved sample load.

D9 *Grass seed*
Grass seed for grassed areas shall comprise the following mix

| | |
|---|---|
| Perennial ryegrass S.23 | 50% |
| Perennial ryegrass S.59 | 30% |
| Smooth stalked meadow grass | 20% |

## Workmanship

D10 *Bottoms of excavations to be approved*
The Contractor shall give the Architect at least two clear working days' notice of when the excavations will be ready for inspection. The bottom of every excavation shall be inspected by the Architect and Building Control Officer and the level agreed between the Architect and the Contractor. No concrete is to be laid until the bottom has been approved and the level recorded.

Any bottom that becomes waterlogged or damaged after approval shall be cleaned out and reformed to the Architect's satisfaction before any concrete is placed.

No concrete or hardcore shall be laid on frozen ground. (Alternatively the excavations shall be trimmed and levelled to 75 mm above the required base depth and, after approval excavated a further 125 mm, trimmed and sealed with a 50 mm layer of blinding.)

D11 *Disposal of excavated material*
All surplus excavated material arising from the Works shall be removed from the site to an approved tip.

D12 *Excavation below required levels*
Should any excavation be taken below the required levels or depths necessary to obtain a suitable bottom, the Contractor shall be required to fill in the excavation to a proper level with material as directed at his own expense.

D13 *Earthwork support*
The Contractor is required to provide all support necessary for securing the sides of trenches and excavations and to be responsible for their safety. Where the Architect instructs or agrees that it is necessary for the safety of the works to retain certain earthwork support, such support shall be measured or agreed before covering up.

D14 *Filling*
Soil for filling to make up levels under floors and pavings and for backfilling around foundations shall be dry, clean subsoil free from clay, topsoil, roots and rubbish. The filling shall be executed in layers not exceeding 200 mm in thickness and shall be well watered and consolidated to ensure proper compaction. Frozen materials or materials containing ice shall not be used for filling.

Filling to make up levels of cultivated areas shall be completed sufficiently in advance of subsequent treatment to allow for natural settlement to take place.

Hardcore shall be consolidated to the required levels and grades with a 2.5 to 3 tonne roller, care being taken not to damage foundations, walls or adjoining work.

D15 *Preparation of cultivated areas*
Cultivation shall be carried out with a mechanical cultivator with fixed tines. All weeds and roots shall be removed and burned and all rubble, large stones and rubbish shall be removed and replaced with topsoil as clause D8.

D16 *Fertilising and grass seeding*
After placing topsoil the cultivated areas shall be dressed with approved fertiliser at the rate of 70 g/m$^2$. The grass seed mixture shall be evenly sown at the rate of 70 g/m$^2$ and covered with 6 mm of fine sifted topsoil and lightly rolled.

## STRUCTURAL STEEL FRAMING
*(SMM Work Section G10)*

## Generally

P1 *Standard of construction*
The whole of the structural steelwork and testing shall comply with the relevant clauses of BS 449 Part 2 'The Use of Structural Steel in Building' and addendum nr 1 and the current edition of The Building Regulations.

P2 *Fabrication by specialist firm*
The steelwork shall be fabricated by a specialist firm approved by the Architect.

P3 *Shop details*
The Contractor shall include for the preparation of all shop details from the drawings supplied by the Architect. These details shall be approved in writing by the Architect before work is put in hand, and shall show full details of all rivets, bolts and welds. The Contractor shall be responsible for the accuracy of his shop details for shop fittings and site connections.

**P4** *Erection arrangements*
The Contractor shall submit to the Architect for information and comment the proposed erection arrangements and programme together with all calculations for erection stresses. Each piece of steel shall be distinctly marked before delivery in accordance with a marking diagram and shall bear such other marks as will assist erection.

## Materials

**P5** *Quality of steel*
   (i) Hot rolled sections, hollow sections, plates, bars and rivets shall comply with BS 449 Part 2 and BS 5950 Part 2.
   (ii) Electrodes for metal arc welding shall comply with BS 639.
   (iii) Black hexagon bolts, screws and nuts shall comply with BS 4190.
   (iv) Precision hexagon bolts, screws and nuts shall comply with BS 3692.
   (v) Black cup and countersunk head bolts and screws with hexagon nuts shall comply with BS 4933.
   (vi) High strength friction grip bolts and associated nuts and washers shall comply with BS 4395 Parts 1, 2 or 3 as described.
   (vii) Plain and taper washers shall be of mild steel.

**P6** *Marking of steel*
Each piece of steel shall be clearly marked with the maker's name or trade mark and with cast numbers or identification marks by which the steel can be traced to the cast from which it was made.

For rivet bars and small pieces securely bundled, a metal tag marked with the cast number will be sufficient.

**P7** *Standard dimensions and weight of steel*
The form, dimensions, weight and tolerance of all rolled shapes and other members shall comply with the British Standards as follows:
   (i) Rolled sections, plates and bars — BS 4 Part 1, BS 5950 or BS 4848.
   (ii) Hollow sections — BS 4848 Part 2.
   (iii) Rivets — BS 4620.
   (iv) Black or precision bolts, screws and nuts — BS 4190, BS 3692 or BS 4933 as described.
   (v) High strength friction grip bolts and associated nuts and washers — BS 4395 Parts 1, 2 or 3 as described.
   (vi) Plain washers—BS 4320 Section 2.

For the purposes of measurement, the weight of mild steel shall be given as BS 648, which will form the basis for the measurement of variations. The weights/m given on the drawings do not include the shelf angles riveted to the webs nor the plates riveted to the flanges of rolled steel joists and other sections.

**P8** *Condition of surfaces*
All surfaces of steelwork shall be clean and free from loose millscale and loose rust.

**P9** *Tests and inspection*
The Architect shall be supplied with manufacturers' mill test certificates as and when required. When directed by the Architect, the Contractor shall deliver samples of structural steel for testing to a named testing station. Should the results of any test be unsatisfactory, the whole consignment of steel represented by the sample shall be rejected and replaced by other material of proper quality at the Contractor's expense. The Architect or his representative shall at all reasonable times have free access to the manufacturer's works.

**P10** *Metallic coatings*
Galvanised steelwork shall comply with BS 729 fully coated with zinc after fabrication by complete immersion in a zinc bath in one operation and excess carefully removed. The finished surfaces shall be clean and uniform.

Zinc-sprayed iron and steel shall comply with BS 2569 Part 1. Before spraying, surfaces shall be blast cleaned to a minimum standard of BS 4232, second quality, with a maximum amplitude of 0.1 mm. The nominal thickness of the zinc coating shall be not less than 0.1 mm and at no point less than 0.075 mm.

**P11** *Paint*
Primer for uncoated structural steelwork which is to be primed at works shall be calcium plumbate priming paint complying with BS 3698 Type A.

For zinc-sprayed surfaces to be primed at works, the pre-treatment primer shall be either a single or two-pack product and the intermediate primer a zinc chromate type, both as recommended by the manufacturer of the subsequent paint system.

Bituminous paint shall be a black bituminous paint complying with BS 3416 Type 1.

## Workmanship

**P12** *Workmanship generally*
The whole of the fabrication and erection of the steelwork shall be carried out in accordance with BS 449 Part 2. The welding of steel to BS 5950 must conform to BS 5135.

Any welder's tests shall be made at the Contractor's expense and shall include the costs of any fees incurred by the Employer for witnessing or making such tests. The right is reserved to make non-destructive tests on the welding to determine whether it conforms to the standards laid down in BS 5135.

P13 *Fabrication*
As much of the fabrication of the steelwork as is reasonably practicable shall be completed in the manufacturer's works. Field connections shall be made in accordance with the approved drawings. The Contractor shall give four days' clear notice of steelwork ready for inspection at the manufacturer's works, to facilitate inspection before delivery.

P14 *Joints and connections*
No variation of the number, type or position of the joints or connections shown on the drawings shall be made without the consent of the Architect. If such consent is desired, the Contractor shall submit detailed drawings of the proposed joints for the approval of the Architect and no extra cost incurred by reason of such additions or alterations will be allowed to the Contractor.

P15 *Painting at works*
Where described as 'primed at works', steelwork shall be free of rust, millscale, welding slag and flux residue and shall be dry immediately prior to painting with primer as clause P11. Zinc-sprayed surfaces shall be primed in accordance with clause P11 as soon as practicable after the zinc coat is applied. Galvanised surfaces to receive bituminous paint shall be clean, dry and free from grease and salts.

For joints with high strength friction grip bolts, the contact surfaces shall be left unpainted but special care shall be taken after assembly to paint all edges and corners near the joints, together with bolt heads, nuts and washers to prevent ingress of moisture.

For joints made with other bolts and rivets, the contact surfaces shall each be given a coat of priming paint and for shop connections the contact surfaces shall be brought together while the paint is still wet.

For welded connections where the contact surfaces are not completely sealed, the contact surfaces shall be painted to within 50 mm of the edges that are to be welded.

The primer shall be touched up with similar paint if damaged by subsequent handling.

## CO-ORDINATED PROJECT INFORMATION

Research by the Building Research Establishment has shown that the biggest single cause of quality problems on building sites is unclear or missing project information. Another significant cause is unco-ordinated design, and on occasions much of the time of site management can be devoted to searching for missing information or reconciling inconsistencies in the data supplied.

The crux of the problem is that for most building projects the total package of information provided to the contractor for tendering and construction is produced in a variety of offices of different disciplines.

To overcome these weaknesses, the Co-ordinating Committee for Project Information (CCPI) was formed on the recommendation of the Project Information Group, sponsored by the following four bodies: Association of Consulting Engineers, Building Employers Confederation, Royal Institute of British Architects and Royal Institution of Chartered Surveyors. Its brief was to clarify, simplify and co-ordinate the national conventions used in the preparation of project documentation.

The following five documents were published either by CCPI or by the separate sponsoring bodies, during 1987 and 1988.

(1) Common Arrangement of Work Sections for Building Works.
(2) Project Specification — a code of procedure for building works.
(3) Production Drawings — a code of procedure for building works.
(4) Bills of Quantities — a code of procedure for building works.
(5) SMM7 (Standard Method of Measurement of Building Works: Seventh Edition).

It is, however, unlikely that any single discipline office will require all these documents. For example, SMM7 conforms to the Common Arrangement and so quantity surveyors using the Standard Method will not need the latter document. Similarly, users of the National Building Specification (NBS) and the National Engineering Specification (NES) will not require the Common Arrangement.

*Common Arrangement of Work Sections for Building Works*
This document plays a major role in co-ordinating the arrangement of drawings, specifications and bills of quantities. It reflects the current pattern of sub-contracting and work organisation in building. To avoid problems of overlap between similar or related work sections, each section contains a comprehensive list of what is included in the

section and what is excluded, stating the appropriate section of the excluded items.

SMM7 uses the same work sections and this will eliminate any inconsistencies between specifications and bills of quantities, where the quantity surveyor structures the bill on SMM7.

The *Common Arrangement* has a hierarchical arrangement in three levels, for example:

Level 1: D Groundwork
Level 2: D3 Piling
Level 3: D30 Cast in place concrete piling (work section).

It lists 24 level 1 group headings and about 300 work sections, roughly equally divided between building fabric and services. However, no single project will encompass more than a relatively small number of them. Only levels 1 and 3 will normally be used in specifications and bills of quantities, while level 2 allows for the insertion of new works sections if required later, without recourse to extensive renumbering.

*Common Arrangement* describes how a work section is a dual concept, involving the resources being used and also the parts of the work being constructed, including their essential functions. The category is usually influenced and characterised by both input and output. For example, an input of brick or block could have an output of walling, while an input of mastic asphalt could have an output of tanking.

Section numbers are kept short for ease of reference. The widespread use of cross references to the specification should encourage designers to be more consistent in the amount of description which they provide on drawings.

*Project Specification — a code of procedure for building works*
This code draws a distinction between specification information and the project specification. Information may be provided on drawings, in bills of quantities or schedules, but project specifications should be the first point of reference when details of the type and quality of materials and work are required. Hence drawings and bills of quantities should identify kinds of work but not specify them. Instead simple cross references should ideally be made to the specification as, for example, Ledkore dampproof course F30.2.

The project specification should be prepared by the designer and the use of a standard library of specification clauses will make the task easier. Specifications should be arranged on the basis of the *Common Arrangement*. The library of clauses in both the National Building Specification and the National Engineering Specification for services installations are so arranged.

The code provides extensive check lists for the specification of each work section to ensure that project specifications are complete, and it also gives advice on specification preparation by reference to British Standards or other published documents or by description.

*Production Drawings — a code for building works*
This code deals with the management of the preparation, co-ordination and issue of sets of drawings, and with the programming of the design and communication operation, and thus complements BS 1192: 1984 (Construction Drawing Practice).

The following criteria should preferably be adopted in the preparation of drawings:

(1) Use of common terminology. If the content of a drawing coincides with a Common Arrangement work section, then the Common Arrangement title should be used on that drawing.
(2) Annotate drawings by cross reference to specification clause numbers, for example, concrete mix A, E.10.4 and lead flashing, H/1.

*SMM7 and Bills of Quantities — a code of procedure for building works*
The arrangement of SMM7 is based on the *Common Arrangement* and the rules of measurement for each work section are in the same sequence.

If descriptions in the bills of quantities are cross referenced to clause numbers in the specification, for example concrete mix A, E10.4, as for the drawings, then the co-ordination of drawings, specifications and bills of quantities will be improved, and the risk of inconsistent information will be reduced. If required the specification can be incorporated into the bill of quantities as preambles.

The code of procedure explains and enlarges upon the SMM as necessary and gives guidance on the arrangement of bills of quantities. It should be emphasised that the code is for guidance only and does not have the mandatory status of SMM7.

*Standard Classifications*
The international SfB system is scheduled to be revised in 1990, and it is possible that the level 1 headings from the *Common Arrangement* may be incorporated in the revised version. The incorporation of the *Common Arrangement*

into CI/SfB, the United Kingdom version of SfB, at its next major revision is under active consideration.

There are also indications that the structure of British Standard specifications may be based on the *Common Arrangement*, so further extending the scope of co-ordinated project information.

*Co-ordinated Project Information in Use*

Since the *Common Arrangement* is based on natural groupings of work within the building industry, it is likely to provide benefits in the management of the construction stages. Not only will it be much easier to find the required information, but it will also be structured by the *Common Arrangement* in a manner which conforms to normal sub-contracting and specialist contracting practice. Thus in obtaining estimates from sub-contractors, it will be a straightforward task to assemble the correct set of drawings, specification clauses and bill items. In management contracting, the *Common Arrangement* is likely to provide a convenient means of identifying separate work packages. Similarly, construction programmes based on the *Common Arrangement* will provide direct links to other project information, thus bringing together quantity, cost and time data into an integrated information package.

Further standardisation has been introduced through the publication of *SMM7 Library of Standard Descriptions*, jointly sponsored by the Property Services Agency (DOE), Royal Institution of Chartered Surveyors and Building Employers Confederation. However, quantity surveyors are not obliged to refer to any of the Co-ordinated Project Information (CPI) documents apart from SMM7 when producing bills of quantities. Hence, in practice, a variety of approaches can be adopted when framing billed descriptions including cross references to project specification clauses, use of the *SMM7 Library of Standard Descriptions*, individual descriptions built up from SMM7 by quantity surveyors as implemented in the worked examples in this textbook, traditional prose given as another acceptable option in the preface to SMM7, or the use of *Shorter Bills of Quantities: The Concise Standard Phraseology and Library of Descriptions*.

The latter publication in two extensive volumes, prepared by two eminent quantity surveying practices, seeks to provide guidelines for the measurement of projects of simpler design and/or construction, where less detailed measurement will suffice, and offers further limited additional simplification of the measurement rules compared with SMM7. It contains very clear classification tables with up to six levels of description with numerous specification examples and an extensive library of standard descriptions. It does not however have the official approval of the SMM7 sponsoring bodies but this is unlikely to preclude its use by some quantity surveying practices.

*Main changes in SMM7*

The Standard method of Measurement is now compatible with the other CCPI publications, as described earlier. Another main change from previous editions of the SMM is that the measurement rules have been translated from prose into classification tables. This approach enables a quicker and more systematic use to be made of the measurement rules and readily lends itself to the use of standard phraseology and computerisation. In addition, the rules have been simplified to produce shorter bills and the document has been updated to conform to modern practice.

The rules prescribed in SMM7 are set out in the form of tables and these comprise classification tables and supplementary rules. Horizontal lines divide the classification table and supplementary rules into zones to which different rules apply. Where broken horizontal lines appear within a classification table, the rules entered above and below these lines may be used as alternatives (SMM General Rules 2.1–3). As for example, metal sheet flashings, aprons, cappings and the like may be measured either with a dimensioned description or with a dimensional diagram (SMM H70.10–18.1–2). Within the supplementary rules everything above the horizontal line, which is immediately below the classification table heading, is applicable throughout that table (SMM General Rules 2.8).

The left-hand column of a classification table lists descriptive features commonly encountered in building works, followed by the relevant unit of measurement. The next or second column lists sub-groups into which each main group shall be divided and the third column provides for further subdivision, although these lists are not intended to be exhaustive. Each item description shall identify the work relating to one descriptive feature drawn from each of the first three columns in the classification table, and as many of the features in the fourth or last column as are appropriate. Where the abbreviation (nr) is given in the classification table, that quantity shall be stated in the item description (SMM General Rules 2.4–7).

The supplementary rules form an extension of the classification tables and are subdivided into the following four columns:

(1) measurement rules prescribe when and how work shall be measured;

(2) definition rules define the extent and limits of the work contained in the rules and subsequently used in the preparation of bills of quantities;

(3) coverage rules draw attention to incidental work which is deemed to be included in appropriate items in the bill of quantities to the extent that such work is included in the project, and where coverage rules include materials they shall be mentioned in item descriptions;

(4) supplementary information contains rules covering any additional information that is required (SMM General Rules 2.9–12).

Cross references within the classification tables encompass the numbers from the four columns, such as D20.2.6.2.0: excavating trenches; width $\leq 0.30$ m; maximum depth $\leq 1.00$ m. The digit 0 indicates that there are no entries in the column in which it appears, while an asterisk represents all entries to the column in which it occurs (SMM General Rules 12.2–4).

A list of symbols and abbreviations is given in SMM General Rules 12.1 and includes m (metre), $m^2$ (sq. m), $m^3$ (cu. m), mm (millimetre), nr (number), t (tonne) hr (hour), $>$ (exceeding), $\leq$ (not exceeding), $<$ (less than), % (percentage) and – (hyphen; often used to denote range of dimensions).

The wording of billed descriptions can vary considerably and it is possible to interpret and implement the provisions of SMM 7 in differing ways. For instance, the author uses the terms thickness and width followed by the appropriate dimensions and excluding the mm symbol, whereas others may prefer to use expressions such as 150 mm thick and 50 mm wide following past practice. The main advantages to be gained by adopting the suggested approach are that it conforms more closely to the wording of SMM7, permits greater rationalisation, facilitates computerisation, and is similar to the method used in the measurement and description of civil engineering work based on the *Civil Engineering Standard Method of Measurement,* thereby securing increased uniformity in the description of measurement of all types of construction work. Similarly, some surveyors may prefer to use the traditional terms 'not exceeding' and 'exceeding' instead of the symbols $\leq$ and $>$. However, these symbols are used throughout SMM7, have the merit of brevity and clarity and will, in the opinion of the author, soon gain general recognition and usage.

It seems likely that there will, in practice, be a variety of different methods adopted for framing billed descriptions, despite the extensive work undertaken by the Building Project Information Committee and the sponsoring bodies, and the wealth of published integrated documentation previously described. It is anticipated that many architects' drawings and specifications will continue to be prepared without reference to the codes of procedure for production drawings and project specifications and the national specifications, and that many quantity surveyors will tend to follow their own personal preferences with regard to bill preparation, so that one universal procedure is unlikely to emerge. Furthermore, the preface to SMM7 permits some flexibility in writing bills of quantities and does not prohibit the use of standard prose. However, the quantity surveyor should never lose sight of the prime objective: namely, to produce bills which fully and accurately represent the quantity and quality of the works to be carried out, founded on a uniform basis for measuring building works emanating from the Standard Method and embodying the essentials of good practice as defined in SMM General Rules 1.1.

## USE OF COMPUTERS IN BILL PRODUCTION

*Computers and Information Technology*
The Royal Institution of Chartered Surveyors was investigating the use of computers as an aid to the preparation of bills of quantities as long ago as 1961. In 1971, the Royal Institution declared that "the quantity surveyor, like others in the industry, must be familiar with and learn to use the computer as it opens up new techniques for the more effective practice of his skills," and in 1983 stated that "computer and information technology can aid professional competence." The 1983 report on *The Future Role of the Chartered Quantity Surveyor* described how rapidly advancing technology in the mini/micro computer field, in word processing and in database information retrieval systems will have a tremendous impact on quantity surveying techniques in the next decade, with much improved services to the client.

These developments will assist in producing more accurate assessments of alternative bids in terms of time and cost valued against higher costs and different client's requirements, such as early completion. In addition, the ability to relate funding, cash flow and ordering of resources will result in more sophisticated methods in the financial management of projects becoming available.

## Microcomputers and bill production

Many makes of microcomputer came on to the market in the 1980s at continually reducing prices. They generally consist of a combined keyboard and monitor (visual display unit or screen), one or two floppy disk (disc) drives and a printer of selected speed and quality. The floppy disks are used extensively as they are relatively cheap and possess quite fast data transfer speeds. The hard disk systems provide increased storage capacity and higher operating speeds. The main disadvantage of microcomputers stems from the incompatability of much of the equipment, although this problem was being reduced in the mid 1980s. The risk of loss or damage is easily overcome by making duplicate copies of all disks.

An increasing use is being made of microcomputers for the preparation of bills of quantities, resulting from the reduced cost of hardware (equipment) and the improved range and efficiency of software (programs). This has culminated in the provision of a more efficient, improved and faster service. The newer models enable more data to be held on the computer with greater ease of access. An increased number of software packages (programs) are available and these permit the production of bills in different formats.

Computer-aided bill production systems provide the facility to check accuracy, but care is needed in the coding of dimensions and entry of data. Modern computerised billing systems can, however, print out errors in the form of tables. The coding can be double checked, although a random check may be considered adequate. The need to engage outside agencies for computerised bill production has been largely eliminated.

## Computerised billing systems

A number of computer systems come within the broad category of the basic bill production type as illustrated in the following figure produced by Adrian Smith in *Computers and Quantity Surveyors*.

The main operations involved in the basic bill preparation process by computer are now listed.

(1) The abstracting and billing stages are computerised, resulting in a substantial reduction in the amount of skilled labour required.

(2) A semi-skilled coding process is used to collect like items together and subsequently to sort them into correct bill sequence. A data entry operation follows in which the coded information is entered into the computer.

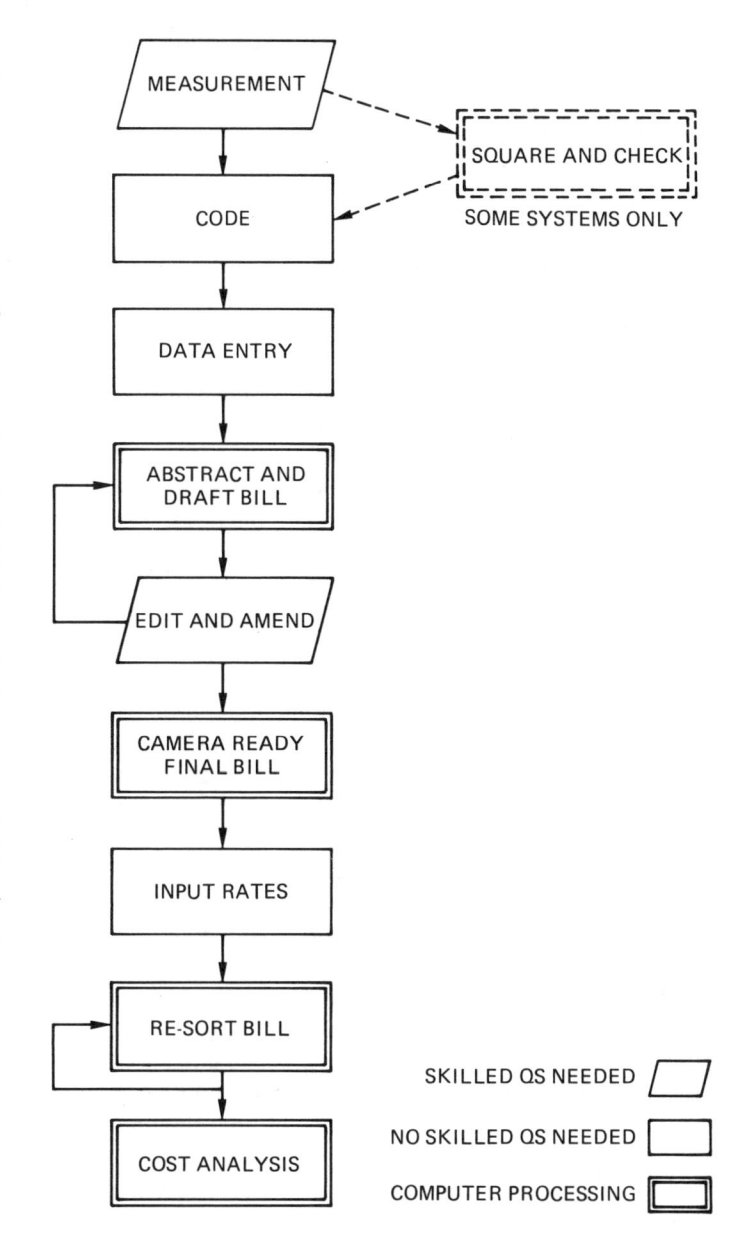

(3) In some systems the dimensions have to be squared prior to the coded information being entered into the computer, whereas others permit the entry of unsquared dimensions when the computer will carry out the necessary calculations.

(4) A feedback loop is incorporated into the process at the abstracting stage, with provision for correction, amendment or alteration of the data, followed by re-processing of the draft bill as often as necessary to ensure complete accuracy before the final printing.

(5) Contractor's tender prices can be added to the information already stored in the computer. Some systems also provide the facility to re-sort the information into alternative bill formats.

(6) Some systems also permit the automatic production of elemental cost analyses from the priced data.

Hence computerised bill production systems provide not only the automated production of the bill, but also subsequent re-use of the data for both pre-contract and post-contract work. The major cost saving is likely to accrue from the computer producing a 'camera ready' final print of the bill ready for duplication, thereby eliminating the costly and time-consuming typing process required in the manual system. It could also show limited cost savings over bill production by cut and shuffle methods.

*Coding and use of libraries of standard descriptions*
It was found necessary to classify each descriptive item with a code, which the computer could then use to collect like items together and to sort the cumulative items into a pre-defined order. The coding operation produced the need for a reference manual for use by the coder, in which each description was listed together with its code, and the manual was generally referred to as a library of standard descriptions.

Most quantity surveying organisations had for a long time produced schedules of standard office descriptions, even if merely derived from descriptions used in previous bills. The next logical step forward was the production of a library of standard descriptions for universal use throughout the construction industry. The principal advantages of industry-wide standard descriptions have been identified as improved quality of written communication, aid to pricing and estimating and simplification of the drafting and understanding of descriptions.

Despite the introduction of a number of libraries of standard descriptions, notably that produced by Fletcher and Moore, many quantity surveyors have been loath to dispense with their own house style with its distinctive prose, but attitudes are likely to change in the future. Smith in *Computers and Quantity Surveyors* has described how a structured library can be developed from SMM7. The publication of the *SMM7 Library of Standard Descriptions* in 1988 could result in a greater awareness of the advantages of using such a document in practice, although its use is optional.

The *SMM7 Library of Standard Descriptions* uses a relatively straightforward mnemonic coding system of abbreviations linked together in the manner shown in the following example.

INSITU CO/REINFCD/SLAB/SPEC B/NETHK 150

comprising the following components:

| | |
|---|---|
| INSITU CO | *In situ* concrete |
| REINFCD | Reinforced |
| SLAB | Slabs |
| SPEC B | Specification B (materials and mix) |
| NETHK 150 | not exceeding 150 thick |

However, the order of items and the terminology used in the *SMM7 Library of Standard Descriptions* does on occasions vary from that adopted in SMM7.

All coded library systems have to make provision for the coding of items not contained in the library. These items are usually termed 'rogue items' and arise on most projects. Dealing with rogue items can be very inconvenient and time-consuming as, in addition to coding the item, the coder usually has to instruct the computer as to the meaning of the code for the rogue item. This process often entails selecting a suitable unused code and telling the computer where to place the item in the bill. The ease or otherwise with which rogue items are accommodated is likely to be a major factor in determining the cost effectiveness of a particular system.

*Operation of coding systems*
The most common approach involves the use of visual display unit (VDU) screens both for coding and the input of data. The computer displays on the screen all the descriptions and their associated codes which are stored in the library and which are relevant to a specific level of item description, and the operator then chooses the appropriate code. It does, in effect, constitute an automated library.

Early systems incorporating this technique required the user to type in the selected code on the keyboard, but more sophisticated methods enable the user to make a selection by merely pointing to the relevant item on the screen, using a cursor, mouse, light pen or other electronic device. This development leads logically to the direct entry of dimensions by the taker off through his personal VDU and the elimination of the separate coding process. A printed copy of the dimensions and the selected descriptions is normally generated simultaneously on a small printer attached to the VDU.

This procedure can be very cost effective since the separate coding process is eliminated although, depending on the response speed of the computer, the taking off time may be increased. Furthermore, in the absence of a centralised data input system, it is necessary for each taker off to have access to a VDU and this can be a problem since the taking off workload in an office is rarely constant, and is not capable of being evenly spread over time. For these reasons each taker off needs his own terminal, resulting in high initial cost, particularly in an office where a large number of takers off are employed, and it can also involve considerable under-use of terminals during slack taking off periods.

An alternative approach is the use of the Phraseology Independent Billing System (PHIBS) developed principally by Derbyshire County Council. This system has no library, leaving the taker off free to use his preferred terminology, provided that the descriptions are properly structured in appropriate levels. The text, as written by the taker off, is entered into the computer by a clerk/typist. The computer scans the text at each level for keywords from which it constructs an internal code. This hidden code is used to collect like items together and to sort the data into bill order. This system simplifies the taking off process in various ways as described by Smith in *Computers and Quantity Surveyors*, including the elimination of coding as a manual process without increasing the workload of the taker off, and by dispensing with the need for each taker off to have constant access to a VDU.

*Electronic/automated measurement*
Systems also exist which enable the computer to be used in assisting the measurement process, either by generating quantities for groups of items from one dimension (automated measurement) or by using some form of electronic measurement device, of which the most common is the digitiser.

A digitiser consists of a flat board normally incorporating a matrix of wires and a pointing device. The board can determine the position of the pointer in both horizontal directions and convey this information to the computer in the form of co-ordinates. It can be used for the direct measurement of floor and ceiling finishes, external works including earthworks, and other appropriate lengths, areas and volumes.

The most commonly used automated measurement technique employs the computer to store pre-measured quantities for specific parcels of work, each of which is given a unique code. The whole parcel may subsequently be included in a new scheme, when it will be multiplied as necessary by quoting the appropriate code and the number required, thereby saving considerable time on repetitive work.

Smith also explains in *Computers and Quantity Surveyors* how the generation of unit quantities can form a useful link between computer billing and architect's computer aided design systems. He illustrates in detail the method of constructing a computer unit quantity library store to accommodate all the coded and measured items required for each type of window contained within a range of windows.

# 2 SUBSTRUCTURES

## MEASUREMENT PROCEDURES

Irrespective of the type of work being measured, there are a number of essential factors that should always be considered and applied by the student to follow good established practice and to assist in improving his examination performance. The principal factors are now listed and briefly described.

(1) Examine the drawings carefully to become familiar with the work and to identify any parts where information is lacking. A query list should be prepared listing the omitted details and in the examination some assumptions as to construction or materials may have to be made. The sequence of measurement should then be determined and this will generally follow the order of construction on the site. It is vital that the order of measurement should be logical and eliminate, as far as practicable, the possibility of omission of items.

(2) Interpret carefully and apply in a sound and logical way the provisions of SMM7 as further amplified by the *Code of Procedure for Measurement of Building Works*. SMM7 (Standard Method of Measurement of Building Works: seventh edition) provides a uniform basis for measuring building works and embodies the essentials of good practice but more detailed information than is required by the document shall be given, where necessary, to define the precise nature and extent of the required work (SMM General Rules 1.1). The prime aim must always be to provide tendering contractors with adequate information from which to compute realistic billed rates. When dealing with unusual forms of construction that are not covered in SMM7, the quantity surveyor should devise his own rules of measurement and include these in a preamble.

(3) Ensure a high standard of accuracy in dimensions, working wherever possible from figured dimensions on the drawings (mainly 1:100 or 1:200 plans, sections and elevations or component drawings), and carefully and neatly recording all preliminary calculations in waste (usually on the right-hand side of the description column). Take care to insert the correct timesing figures on all occasions. Where work cannot be adequately described and measured, it shall be given as a provisional sum, and identified as either defined or undefined work. Where the work can be described but the quantity cannot be accurately determined, an estimate shall be given of the quantity and described as approximate (SMM General Rules 10.1–6).

(4) Make full use of sub-headings to act as signposts throughout the dimensions to facilitate identification and of waste for preliminary calculations, explanatory notes and locational descriptions. Students should aim to make the take off as comprehensive as possible and this will assist enormously in the adjustment of any subsequent variations to the work.

(5) All descriptions should be adequate, concise and written in a logical and easily understood format following the sequence adopted in SMM7 and illustrated in the worked examples in this book. Where it is not possible adequately to describe a billed item, SMM7 makes provision for the inclusion of dimensioned diagrams to assist the contractor in pricing the item (SMM General Rules 5.3). Typical examples of dimensioned diagrams are stone members, such as lintels, sills, mullions and transoms; carpentry fittings, such as straps, hangers, shoes and metal connectors; doors and windows. Dimensioned diagrams shall show the shape and dimensions of the work (SMM General Rules 5.3), whereas component drawings shall show the information necessary for the manufacture and assembly of a component, such as a staircase (SMM General Rules 5.2).

(6) Adopt a logical sequence of items so that each one follows the preceding one in a natural progression with reduced risk of error. It is good practice to list the major

items of work in the proposed taking off sequence prior to starting the actual measurement process and before the student becomes too enmeshed with specific details.

(7) Seek to achieve a good standard of presentation; this also helps to secure extra marks in the examination. Nothing is more off-putting to an examiner than an untidy and ragged set of dimensions with written descriptions that are barely legible. Even in the heat of examination care must be taken to produce good, clearly and orderly presented work with ample space between each entry.

## EXCAVATING AND FILLING

*Generally*

The rules contained in SMM7 are based on mechanical excavation. The *Code of Procedure* states that the information provided in accompanying descriptions should identify those circumstances where it may be difficult or impractical to carry out excavation by mechanical means.

SMM D20.P1 requires information on groundwater levels, trial pits or boreholes and similar information to be shown on location drawings or further drawings accompanying the bill of quantities, or stated as assumed.

The groundwater level and the date when it was established, termed the pre-contract water level, shall be given. (Groundwater is given as a single word in British Standard publications, but SMM7 divides it into two.) The groundwater level is re-established at the time each excavation is carried out (post contract water level) and the appropriate quantities of work below groundwater can then be adjusted where necessary.

Particulars of trial pits or boreholes shall be given in order that the contractor will know the type of soil to be excavated. The positions are normally shown on the location drawings. In the absence of this information, a description of the assumed ground and strata is to be stated to provide a basis from which the contractor can compute his excavation rates.

*Site preparation*

Billed items for the site preparation work of removing trees and tree stumps (enumerated in the girth stages given in SMM D20.1.1–2.1–3.0) and clearing site vegetation, which includes bushes, scrub, undergrowth, hedges, and trees and tree stumps ≤ 600 mm girth, (in m$^2$), are deemed to include grubbing up roots, disposal of materials and filling voids (SMM D20.C1). Lifting turf for preservation is measured in m$^2$, stating the method of preservation (SMM D20.1.4.1.0), while excavating topsoil for preservation is measured in m$^2$, stating the average depth (SMM D20.2.1.1.0).

*Excavating*

The volume of excavation, excluding topsoil for preservation, is measured in m$^3$ as the bulk before excavating, and the estimator has to allow in his prices for the subsequent variations to bulk or for extra space for working space or to accommodate earthwork support (SMM D20.M3). A wide range of excavation classifications is given in SMM D20.2.2–8.1–4.1, such as to reduce levels, basements and the like, pits, trenches (width ≤ and > 0.30 m), for pile caps and ground beams between piles, and to bench sloping ground to receive filling, all in the prescribed maximum depth range stages (≤ 0.25 m, 1.00 m, 2.00 m and thereafter in 2.00 m stages). Separate items as extra over any types of excavating irrespective of depth are required for excavating below groundwater level (m$^3$), next existing services (m), and around existing services crossing excavation (nr), stating the type of service, such as water service.

Breaking out existing materials is measured in m$^3$ as extra over any types of excavating irrespective of depth and classified in one of the categories listed in SMM D20.4.0.1–5.1, that is, rock; concrete; reinforced concrete; brickwork, blockwork or stonework; and coated macadam or asphalt. Breaking out existing hard pavings are measured as extra over items in m$^2$, stating the thickness (SMM D20.5.0.2, 5.1).

Working space allowance to excavations is classified in one of the four categories listed in SMM D20.6.1–4.0.0, and measured where the face of the excavation is < 600 mm from the face of formwork, rendering, tanking or protective walls (SMM D20.M7). The area measured is calculated by multiplying the girth of the formwork, rendering, tanking or protective walls by the depth of excavation below the commencing level of the excavation (SMM D20.M8). Additional earthwork support, disposal, backfilling, work below groundwater level and breaking out are deemed to be included (SMM D20.C2).

*Earthwork support*

There is no requirement to separate earthwork support to different types of excavation. Earthwork support includes the use of timber planking and strutting, plywood trench sheeting and light steel trench sheeting and strutting. Interlocking driven sheet piling is excluded from the earthwork support classification and where required by the design or specification shall be measured in accordance with Section D32.

Support to the sides of excavations is measured in m² to trenches (excluding pipe trenches where it is deemed to be included in the linear trench excavation item), pits, and the like, where > 0.25 m deep, whether the support will actually be required on the site or not. The maximum depth is given in stages in accordance with SMM D20.7.1–3. Earthwork is also classified by the distance between opposing faces in stages of ≤ 2.00 m, 2.00 m to 4.00 m, and > 4.00 m. Where the opposing face occurs on the opposite side of a building, the distance between opposing faces will fall into the > 4.00 m category, and the contractor will have to decide whether raking struts are required. Earthwork support left in, curved, next to roadways or existing buildings, below groundwater level or to unstable ground shall be so described and measured separately in accordance with the rules contained in SMM D20.6–8 and as illustrated in the *Code of Procedure*.

In the case of excavations designed with set-backs, each vertical face between set-backs should be considered separately. However, this would not apply where the contractor, at his discretion, decided to use set-backs, probably as an aid to earthwork support.

*Disposal of water and excavated material*
A billed item is required for disposal of surface water, that is, rainwater as distinct from groundwater (SMM D20.8.1.0.0), with a further item where groundwater disposal is required (SMM D20.8.2.0.0). An adjustment of the billed amount for the latter will be needed if the pre-contract and post-contract water levels differ.

The subsequent disposal of excavated material forms a separate billed item in m³, either of soil to be stored on site, used as filling to make up levels, filling to excavations, or to be removed off the site. In the first instance, when measuring trench excavation, it is usual to take the full volume as filling to excavations and subsequently to adjust as disposal of excavated materials off site with the measurement of concrete and brickwork, as illustrated in Example 1. With basement excavation it will normally be more convenient to measure all the excavated soil for disposal from the outset.

Unless there are specific requirements as to the handling or disposal locations, the handling of the excavated material will be at the discretion of the contractor. For example, where excavated material is to be disposed on site, the description should include any specific requirements for the location of such deposits and the average distance from the excavation in metres.

*Filling*
Filling is measured in m³ in one of the three categories prescribed in SMM D20.9–11, and separating that with an average thickness ≤ or > 0.25 m. The student should note the three classifications for sources of filling material in SMM D20.9–11.1–2.1–3, namely, arising from excavations, obtained from on site spoil heaps, and obtained off site stating the type of material. Further details may be included in the bill description, where appropriate, such as topsoil, or selected, treated or specified handling details, as listed in the fourth column of SMM D20.9–11.

*Surface treatments*
The measurement of the excavation is generally followed by a superficial item for compacting the bottom of the excavation (SMM D20.13.2.3.0). Surface treatments may alternatively be given in the description of any superficial item (SMM D20.M17). Compacting is deemed to include levelling and grading to falls and slopes ≤ 15° from horizontal (SMM D20.C5). Compacting ground, filling and bottoms of excavations each generate separate items, and where the surface to be compacted is blinded, the nature of the blinding material shall be stated. Other surface treatments listed in SMM D20.13.1–5 include applying herbicides, trimming (sloping surfaces, sides of cuttings and sides of embankments), trimming rock to produce a fair or exposed face, and preparing subsoil for topsoil.

## SLOPING SITE EXCAVATION AND ASSOCIATED WORKS

Worked examples of the measurement of excavation work for strip foundations to a small building and a curved brick screen wall and a basement were included in *Building Quantities Explained*. Students are advised to revise this work before proceeding to study example 1 in this chapter covering stepped foundations to a small garage block.

On a sloping site it is necessary to determine, from the ground levels or contours, which parts of the site require reduced level excavation to reach formation level and which parts require fill. The boundary between the two (where the stripped level coincides with the formation level) is sometimes termed the 'cut and fill' line. Further lines are drawn to show the limit of excavation not exceeding 0.25 m deep, another at not exceeding 1.00 m deep and continuing in the maximum depth stages given in SMM D20.2.1.1–4.1 until all the excavation work is covered.

With the fill, one dividing line only is needed at a depth of 250 mm. The positions of these demarcation lines may be determined by scaling or by calculation and they are frequently irregular in outline.

The excavation to foundation trenches of varying depths will have to be separated into appropriate lengths within the prescribed maximum depth ranges. In example 1 all the depths fall within the 1.00 m range and hence the lengths on the front and rear walls are determined by the steps and the average depths for the excavation dimensions obtained by calculation. Following normal practice the trench excavation volume is also taken as fill in the first instance to be subsequently adjusted for the space occupied by concrete and brickwork.

Methodically work through compacting of bottoms of excavations, earthwork support to sides of trenches with adjustments at trench intersections, concrete with adjustments for steps and the surface water disposal item. The remainder of the work up to damp-proof course then follows with the sections of brickwork on the front and rear walls of varying depths not coinciding with the excavation lengths. Note how a checking process is applied to ensure that the full girth is taken for each class of work. Some projections have been included to show the method of measurement of the various component parts. Remember to fill the void outside the enclosing walls with topsoil.

The example is completed by measuring the filling, hardcore, damp-proof membrane and concrete floor slab. Students are advised to work carefully through the dimensions, paying particular attention to the explanatory notes and checking back on the references to SMM7. As a further aid to study, the student is advised to make a list of the appropriate items in the correct sequence, without entering dimensions, by reference to the drawing but without looking at the dimensions in the book.

## PILING

Piling work is covered in three work sections of SMM7, namely D30: Cast in place concrete piling; D31: Preformed concrete piling; and D32: Steel piling. General information to be provided is common to all three work sections in P1–3, as now described. The accompanying location drawings are required to show the general piling layout; the positions of different types of piles; the positions of the work within the site and of existing services; and the relationship to adjoining buildings. A description shall be provided of the nature of the ground and, where the work is carried out near canals, rivers or tidal waters, the levels of the ground shall be given in relation to the relevant water levels. The levels from which the work is expected to begin and from which measurements have been taken are stated, and irregular ground is so described. In all cases driven depths are measured from the commencing surface to the bottom of the pile.

*Cast in place concrete piles*

Bored and driven shell piles shall include the nominal diameter and are made up of three components: (1) total number stating the commencing surface; (2) total concreted length in metres; and (3) total length in metres, stating the maximum depth. Preliminary piles, contiguous bored piles and raking piles are separate items (SMM D30.1–2.1.1–3.1–3). Items are taken extra over piling for breaking through obstructions by the hour, and enlarging bases for bored and driven piles by number, stating the diameter of the enlarged base (SMM D30.5.1–3.1.0).

Additional items comprise permanent casings in metres, cutting off tops of piles (nr) in metres, reinforcement in tonnes, disposal of excavated materials in m$^3$, delays with rig standing, where specifically authorised, by the hour, and pile tests by number stating details of the tests (SMM D30.6–11). Cutting off tops of piles is deemed to include preparation and integration of reinforcement into pile cap or ground beam and disposal (SMM D30.C5), while reinforcement to piles is deemed to include tying wire, spacers, links and binders which are at the discretion of the contractor (SMM D30.C6).

*Preformed concrete piling*

Reinforced, prestressed, reinforced sheet and hollow section piles are each measured separately, stating the nominal cross-sectional size in each case. There are two measured items, namely: the total number driven stating the specified length and commencing surface and the total driven depth in metres, with preliminary and raking piles separately classified (SMM D31.1–4.1.1–2.1–2). Driving heads and shoes are deemed to be included (SMM D31.C1).

Filling hollow piles with concrete is measured in metres, distinguishing between plain and reinforced concrete. Pile extensions are enumerated and extension lengths given in metres, separating those ≤ and > 3.00 m (SMM D31.8–9). Similar items to those described for cast in place concrete piling are taken for cutting off tops of piles, disposal of excavated material, delays and pile tests (SMM D31.10–13).

The author has doubts about the appropriateness of the method prescribed for the measurement of disposal of excavated material for this class of pile, and it would probably have been better for the items for piles to be deemed to include disposal of excavated material, as in the *Civil Engineering Standard Method of Measurement*, to overcome the problem.

*Steel piling*
Descriptions of isolated piles are to include the mass per metre and cross-sectional size or section reference. There are two component items, namely: the total number driven giving the specified length and commencing surface, and the total driven depth in metres, separating preliminary piles, raking piles and those to be extracted (SMM D32.1.1.1–2.1–3).

Interlocking piles are described by section modulus and cross-sectional size or section reference. The total area of specified length is measured in m$^2$, separating those with a length ≤ 14.00 m, 14.00–24.00 m, and those > 24.00 m, and the total driven area is given as a separate item (SMM D32.2.1.1–4.0). The areas of interlocking piles are calculated by multiplying the mean undeveloped horizontal lengths of the pile walls (including lengths occupied by special piles) by the appropriate depths or lengths (SMM D32.M4).

Corners, junctions, closures and tapers are measured in metres as extra over interlocking piles (SMM D32.3.1–4.1.0). Pile extensions, delays and pile tests are measured as previously described for the other classes of pile. Cutting off surplus pile from specified lengths is measured in metres stating the number involved, and distinguishing between isolated and interlocking piles and between preliminary and raking piles (SMM D32.6.1–2.1–2.1–2).

Example 2 provides worked examples of bored cast in place concrete piles, preformed concrete piles and interlocking steel piles to give a reasonable coverage of this class of work.

## UNDERPINNING

General information as prescribed in SMM D50.P1–3, is to be provided in the underpinning work section, and this embraces location or further drawings showing the location and extent of the work and details of the existing structure to be underpinned and, in the case of a complicated design, detailed drawings may be required. Information will also be required relating to the excavation work and in difficult ground conditions full particulars should be given of special requirements to be taken. Details are to be supplied of the limit of length to be carried out in one operation and the number of sections the contractor is permitted to undertake at one time.

Temporary support for existing structures is given as an enumerated item, stating any particular requirements, including details of making good (SMM D50.1.1.0.0). Excavating work is subdivided into preliminary trenches and underpinning pits both measured in m$^3$ to the normal maximum depth classifications and separating work which is curved, undertaken from one side only or from both sides (SMM D50.2.1–2.1–4.1–3). Preliminary trenches extend down to the underside of existing foundations (SMM D50.D1), while underpinning pits extend from the underside of existing foundations down to the base of the underpinning excavation (SMM D50.D2).

Width allowances are related to the total depth of excavation measured from the top of the preliminary trench to the base of the underpinning pit as follows:

(a) 1 m where the total depth is    1.5 m
(b) 1.5 m where the total depth is  1.5–3 m
(c) 2 m where the total depth is >  3 m (SMM D50.M1).

The width of a preliminary trench is calculated as the sum of any projection of the retained foundation beyond the face of the wall plus any projection of the underpinning beyond the face of the retained foundation plus the width allowance (SMM D50.M2). When calculating the width of an underpinning pit, use the preliminary trench approach but substitute 'the sum of the width of retained foundation' for 'the sum of any projection of the retained foundation beyond the face of the wall' (SMM D50.M3).

Items extra over any type of excavating are measured in accordance with SMM D20.3–5. Earthwork support to preliminary trenches is kept separate from that to underpinning pits and measured in accordance with SMM D20.7. Earthwork support to underpinning pits is measured to the back, front and both ends of the underpinning pits and also between each section of the underpinning (SMM D50.M6).

Cutting away existing projecting foundations is measured in metres, stating the maximum width and depth of projection and distinguishing between masonry and concrete (SMM D50.5.1–2.1.0). Preparing the underside of the existing work to receive the pinning up of the new work is also measured in metres, stating the width of the existing work (SMM D50.6.1.0.0).

Disposal of water and excavated material, filling, and surface treatments are measured in accordance with SMM D20, as described earlier in the chapter. Concrete, formwork, brickwork and tanking are measured in accordance with the appropriate work sections (SMM D50.10–14).

The worked example of underpinning work in example 3 will serve to illustrate the application of the prescribed rules.

# FOR READER'S NOTES

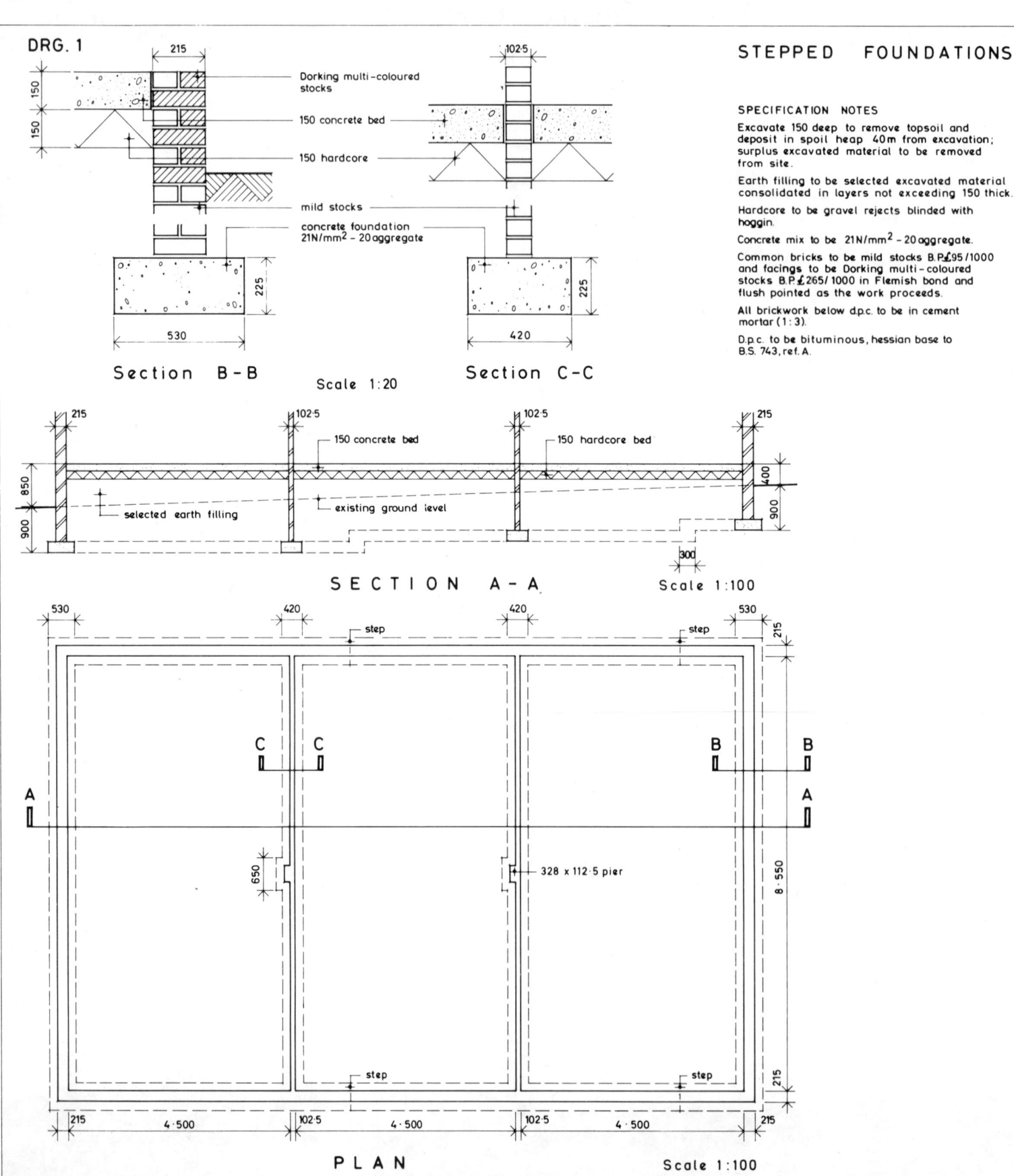

# STEPPED FOUNDATIONS

## SPECIFICATION NOTES

Excavate 150 deep to remove topsoil and deposit in spoil heap 40m from excavation; surplus excavated material to be removed from site.

Earth filling to be selected excavated material consolidated in layers not exceeding 150 thick.

Hardcore to be gravel rejects blinded with hoggin.

Concrete mix to be 21N/mm$^2$ – 20 aggregate.

Common bricks to be mild stocks B.P. £95/1000 and facings to be Dorking multi-coloured stocks B.P. £265/1000 in Flemish bond and flush pointed as the work proceeds.

All brickwork below d.p.c. to be in cement mortar (1:3).

D.p.c. to be bituminous, hessian base to B.S. 743, ref. A.

## STEPPED FOUNDATIONS

| | | | | | |
|---|---|---|---|---|---|
| | The work comprises the sub-structural wk. to a block of 3 garages, measuring 14·135 × 8·980 m o/a on an unobstructed urban site adjoing. a tarmac pub. highway. <br> The ground shall be assumed to consist of 150 mm of topsoil overlaying sandy gravel. <br> Groundwater lev. was established as 3·800 m below e.g.l. (153·50) on 15th June 1989. <br> No over grd. or undergrd. services cross the site. | A general description of the work is to be given where it is not evident from the location drawings, mainly for the benefit of sub-contractors (SMM A13). In this example most of the information is evident from the drawing, and so the description is very brief. <br> In the absence of trial pits or boreholes, a description of the ground and strata which is to be assumed shall be stated (SMM D20.P1). <br> Particulars are also required of the groundwater level and the date when it was established in accordance with SMM D20. P1a. <br> This is the pre-contract water level and will be re-established when work is commenced (post contract water level) and the quantities of excavation below groundwater level (extra over item) adjusted where necessary. <br> Information is required about any existing services on the site (SMM D20.P1f). <br> All relevant plant items are covered in the Preliminaries/ General Conditions Bill (SMM A43). | | 14·45 <br> 9·30 <br><br> 14·45 <br> 9·30 <br> 0·15 | Exc. topsoil for preservn. av. 150 dp. <br><br><br> Disposal of excvtd. mat. on site in spoil heaps av. dist. of 40 m from excavn. <br><br> Av. longtdnl. trench depths <br>     middle     upper <br>     sectn.      end <br>     1·100      1·000 <br> less step   225   less step   225 <br>      875          775 <br>     1·000       900 <br>   2) 1·875    2) 1·675 <br>   av.   938    av.   838 <br> less topsoil 150   less topsoil 150 <br>      788          688 | # EXAMPLE 1 <br> Excavating topsoil to be preserved is measured in m² (SMM D20.2.1.1.0) <br><br> Measured in m³ in accordance with SMM D20.8.3.2.1. If the topsoil is subsequently to be spread on site, a further item will be required in m³ stating whether the average thickness is ≤ or > 0·25 m (SMM D20. 10.1-2.2.3). <br> The average depths of each longitudinal stepped section are calculated in waste by scaling the intermediate depths, and adjusting for the steps in the foundation and for the topsoil which has already been measured. It is assumed for the purposes of this example that the site is level across the depth of the garages. |

            Site Prepn.

    fdn. sprd. to ext. walls
                530
                215
          2) 315
            158

        o/a. len.
ext. walls 2/215      430
2ce fdn. sprd. 2/½/315   315
int. walls 2/102·5    205
int. spans 3/4·500   13·500
                  14·450

        o/a. width.
ext. walls 2/215      430
fdn. sprd. 2/½/315    315
int. dimsn.         8·550
                  9·295

No part of the site is above formation level and hence there is no reduced level excavation.
All preliminary calculations must be shown in waste, no matter how insignificant they may appear. Examiners always award marks for logical, accurate and well presented waste calculations. Also insert ample sub-headings to act as signposts throughout the dimensions.
The overall dimensions of topsoil excavation are calculated in waste from the figured dimensions on the drawing.

             lower
             end
            900
          1·100
        2) 2·000
          1·000
less topsoil   150
            850

Longtdl. trench lens.
  scaled upper len.   1·350
  scaled lower len.   6·375
                 7·725
  total len.     14·450
less upper & lower lens. 7·725
  middle len.    6·725

| | | | |
|---|---|---|---|
| 2/ | 6·38 <br> 0·53 <br> 0·85 | Exc. tr. width > 0·30 m, max. depth ≤ 1·00 m. <br> (lower sectn. <br> & | In the absence of figured dimensions it is necessary to scale the lengths of the sections of trench. To avoid cumulative errors and also to provide a check, the combined scaled lengths of the upper and lower sections are deducted from the overall calculated length to give the length of the middle section, which is then scaled as a final double check. The lengths are twiced for both sides of the building. <br> Excavation to foundation trenches is measured in m³ and classified in the prescribed maximum depth stages, and stating the trench width classification and commencing level where > 0·25 m below existing ground level (SMM D20.2.6.1-4.1). <br> All excavated soil from trenches is taken as 'filling to excavations' in the first instance, stating whether the average thickness is ≤ or > 0·25 m and the origin of the filling material (SMM D20.9.2.1.0). |
| 2/ | 6·73 <br> 0·53 <br> 0·79 | Fillg. to excavns., av. thickness > 0·25 m, arisg. from excavns.   (middle sectn. | |
| 2/ | 1·35 <br> 0·53 <br> 0·69 | (upper sectn. | |

1·1                                                                1·2

## STEPPED FOUNDATIONS (Contd.)

### Trenches to cross walls

```
                              len.
o/a. len.                    9.295
less fdn. trs. to frt. &
rear walls      2/530        1.060
                             8.235
                             depth
                            end walls
                              900
              less topsoil   150
                              750
                          dividg. walls
                  1.000        850
              less topsoil  150  150
                              850  700
```

| | | |
|---|---|---|
| 2/ | 8.24<br>0.53<br>0.75 | Exc. tr. a.b.   (end walls<br>&<br> |
| | 8.24<br>0.42<br>0.85 | Fillg. to excavns. a.b.<br>(dividg. wall |
| | 8.24<br>0.42<br>0.70 | (do. |

The lengths of the foundations to the cross walls and their depths are calculated in waste. The lengths are obtained from the overall length previously calculated less the trench width to the front and rear walls.

As the descriptions of these two items are identical to those previously taken; abbreviated descriptions followed by the letters a.b. (as before) are sufficient. The quantities for excavating and subsequent disposal shall be the bulk before excavating (SMM D20.M3) and the estimator is to allow for this in his rates.

### Projs. to dividg. walls.

| | | |
|---|---|---|
| | 0.65<br>0.11<br>0.85 | Exc. tr. a.b.<br>& |
| | 0.65<br>0.11<br>0.70 | Fillg. to excavns. a.b. |

The additional excavation and fill to accommodate the projecting foundations to the attached piers on the dividing walls are now taken to complete the foundation trench excavation.

```
                14.450
                 8.235
              2/22.685
                45.370
```

| | | |
|---|---|---|
| | 45.37<br>0.53 | Compactg. bott. of excavns.<br>(main enclosg. walls |
| 2/ | 8.24<br>0.42 | (dividg. walls |
| 2/ | 0.65<br>0.11 | (projs. to<br>dividg. walls |
| 2/2/ | 6.38<br>0.85 | Earthwk. suppt.<br>max. depth ≤ (long. walls<br>1.00 m, dist. lower sectn.<br>between opposg. |
| 2/2/ | 6.73<br>0.79 | faces ≤ 2.00 m. (long. walls<br>middle sectn. |

Surface treatment of bottoms of excavations is measured in m² (SMM D20.13.2.3.0). Give locations of each section of the work in waste for ease of identification.

Earthwork support is measured in m² stating the maximum depth range and the distance between opposing faces in accordance with the classifications listed in SMM D20.7.1.1.0.

1 · 3

| | | |
|---|---|---|
| 2/2/ | 1.35<br>0.69 | Earthwk. suppt.<br>a.b. (long. walls<br>upper sectn. |
| 2/2/ | 8.24<br>0.75 | (end walls |
| 2/ | 8.24<br>0.85 | (dividg. wall |
| 2/ | 8.24<br>0.70 | (do. |
| 2/ | 0.11<br>0.85 | (proj. to<br>dividg. wall |
| 2/ | 0.11<br>0.70 | (do. |

```
              14.450
               9.295
           2/ 23.745
              47.490
```

| | | |
|---|---|---|
| | 47.49<br>0.15 | Earthwk. suppt. a.b.<br>(exposed face<br>of topsoil |
| 2/ | 0.42<br>0.85 | Ddt. do.<br>(intl. & ext.<br>wall inter-<br>sectns. |
| 2/ | 0.42<br>0.70 | |
| | 45.37<br>0.53<br>0.23 | In situ conc. fdns.<br>(21N/mm² - 20 agg.)<br>poured on or (main enclosg.<br>against earth. walls<br>& |
| 2/ | 8.24<br>0.42<br>0.23 | Ddt Fillg. to (dividg. walls<br>excavns. a.b.<br>& |
| 2/ | 0.65<br>0.10<br>0.23 | Add Disposal of (projs. to<br>excvtd. mat. off (dividg. walls<br>site. |
| 4/ | 0.30<br>0.53<br>0.23 | (steps in fdns. |
| 4/ | 0.53 | Fwk. to sides of fdns. plain<br>vert., ht. ≤ 250. |

Earthwork support is not measured to faces ≤ 0.25 m high (SMM D20.M9a). The earthwork support to the outer face of the trench will include the depth occupied by topsoil as provided in the next item. This can be strutted from the opposing face of the trench and so the distance of ≤ 2.00m can apply.

This item picks up the earthwork support to the exposed face of the topsoil on the outer face of the main foundation trenches.
The earthwork support at the corners has been self-adjusting and requires no further items, but it is necessary to make deductions for the support already measured at the intersection of external and internal walls.

The mix or strength requirements of the concrete shall be stated (SMM E10.S1) and any appropriate classifications included such as foundations and poured on or against earth (SMM E10.1.0.0.5).
Care must be taken to include incidental items such as the foundations to projections to dividing walls and steps in foundations.
The concrete is followed by the adjustment of soil disposal.

The formwork to the face of the steps is measured as a linear item, giving the appropriate classification and height range as SMM E20.1.1.2.0.

1 · 4

## STEPPED FOUNDATIONS (Contd.)

| | | | | | | | |
|---|---|---|---|---|---|---|---|
| Item | | Disposal of surf. water. | As required by SMM D20.8.1.0.0. Excavations are above groundwater level and so the provisions of SMM D20.8.2.0.0 do not apply. The lengths of front and rear walls of varying heights are calculated in waste, by making adjustments from the concrete dimensions for foundation spread and steps. | | 8·55<br>1·53<br>8·55<br>1·30 | Bk. wall, thickness: 102·5, mild stocks a.b. in stret. bond in c.m. (1:3). | (dividg. wall lower end)<br><br>(dividg. wall upper end) | The lengths inserted for the end walls are the internal dimensions as the corners have already been included with the front and rear walls.<br>Dividing walls are then measured up to damp-proof course level. |

Bwk. longtdnl. walls
| | lens. |
|---|---|
| o/a. len. of fdn. | 14·450 |
| less twice fdn. sprd. | 315 |
| | 14·135 |
| | lower sectn. |
| less fdn. sprd. 158 | 6·375 |
| steps 300 | 458 |
| | 5·917 |
| | upper sectn. |
| | 1·350 |
| add step | 300 |
| | 1·650 |
| less fdn. sprd. | 157 |
| | 1·493 |

| lower sectn. | 5·917 |
|---|---|
| middle sectn. | 6·725 |
| upper sectn. | 1·493 |
| total len. | 14·135 |

A final check ensures that the cumulative total length is correct.
The heights of brickwork on the front and rear walls are also calculated in waste, by taking the figured dimensions from the drawing and adjusting for the depths of foundations and steps, and working through logically from one section to the next. The height of the last section is checked against the figured dimensions, thus:

Bwk. hts.
| | lower sectn. |
|---|---|
| | 850 |
| | 900 |
| | 1·750 |
| less fdn. | 225 |
| | 1·525 |
| | middle sectn. |
| | 1·525 |
| less step | 225 |
| | 1·300 |
| | upper sectn. |
| | 1·300 |
| less step | 225 |
| | 1·075 |

| above ground | 400 |
|---|---|
| below ground | 900 |
| | 1·300 |
| less foundations | 225 |
| ht. of brickwork | 1·075 |

| 2/ | 5·92 | | Bk. wall, thickness: 215, mild stocks B.P. £95/1000 (delvd. to site) in Flem. bond in c.m. (1:3). | (long. walls lower sectn.) |
|---|---|---|---|---|
| | 1·53 | | | |
| 2/ | 6·73 | | | (long. walls middle sectn.) |
| | 1·30 | | | |
| 2/ | 1·49 | | | (long. walls upper sectn.) |
| | 1·08 | | | |
| | 8·55 | | | (end wall lower end) |
| | 1·53 | | | |
| | 8·55 | | | (end wall upper end) |
| | 1·08 | | | |

Brickwork is measured in m² stating the thickness, (which could be expressed as 215 or 1B) and classified as in walls (SMM F.10.1.1.1.0). Work is deemed vertical unless otherwise described (SMM F10.D3). The description is to include the kind, quality and size of bricks, bond and composition and mix of mortar (SMM F10.S1-3). Brick sizes are best given in a preamble clause. The longitudinal wall sections are twiced for the front and rear walls.

1·5

Proj., width: 328, depth of proj. 112·5, vert., mild stocks a.b. in Eng. bond, in c.m. (1:3).

(attchd. piers)

Adjust. of excavn. bwk. depths
longtdl. walls
| | lower | middle | upper |
|---|---|---|---|
| | 850 | 788 | 688 |
| less fdn. | 225 | 225 | 225 |
| | 625 | 563 | 463 |

cross walls
| | end | l. divg. | u. divg. |
|---|---|---|---|
| | 750 | 850 | 700 |
| less fdn. | 225 | 225 | 225 |
| | 525 | 625 | 475 |

| 2/ | 5·92 | | Ddt. Fillg. to excavns. a.b.<br>&<br>Add Disposal of excvtd. mat. off site. | |
|---|---|---|---|---|
| | 0·22 | | | |
| | 0·63 | | | (long. walls lower sectn.) |
| 2/ | 6·73 | | | |
| | 0·22 | | | |
| | 0·56 | | | (long. walls middle sectn.) |
| 2/ | 1·49 | | | |
| | 0·22 | | | |
| | 0·46 | | | (long. walls upper sectn.) |
| 2/ | 8·55 | | | |
| | 0·22 | | | |
| | 0·53 | | | (end walls) |
| | 8·55 | | | |
| | 0·10 | | | |
| | 0·63 | | | (lower dividg. wall) |
| | 8·55 | | | |
| | 0·10 | | | |
| | 0·48 | | | (upper dividg. wall) |
| | 0·65 | | | |
| | 0·11 | | | |
| | 0·63 | | | (proj. on lower dividg. wall) |
| | 0·65 | | | |
| | 0·11 | | | |
| | 0·48 | | | (proj. on upper dividg. wall) |

1·6

Projections as in attached piers, plinths, oversailing courses and the like are grouped together and measured in metres, stating the width and depth of the projection and plane (SMM F10.5.1.1.0). English bond has been chosen for maximum strength.
To be classified as projections, length on plan is ≤ 4 times the thickness (SMM F10.D9). It is logical to follow the measurement of the brickwork with the adjustment of the soil disposal for the volume occupied by the brickwork.
These must be cubic items, although examination candidates in haste sometimes enter them as superficial ones.
Note the method of tabulation of the calculations in a compact and orderly way. The initial depths are taken from the foundation excavations, which have already been adjusted for topsoil removal. The concrete foundation depth is deducted as this item has already been adjusted. The lengths are taken directly from the brickwork items. The same sequence of items is adopted throughout, namely front and rear walls, end walls, internal dividing walls and projections to dividing walls. Locational notes are inserted in waste (the right hand side of the description column).

## STEPPED FOUNDATIONS (Contd)

|   |   |   |   |
|---|---|---|---|
| | | Faced bwk. | |
| | | | ht. |
| | | lower corners 2/850 | 1·700 |
| | | upper corners 2/400 | 800 |
| | | mid-points 2/650 | 1·300 |
| | | 6) | 3·800 |
| | | av. ht. above g.l. | 633 |
| | | add one course below g.l. | 75 |
| | | av. ht. of facgs. below dpc | 708 |

| 45·37 | | Add Bk. wall, facewk. o.s., thickness: 215, Dorking multi-coloured stocks, B.P. £ 265/1000 (delvd. to site) in Flem. bond & ptg. in c.m. (1:3) & ptg. w. nt. flush jt. as wk. proceeds. (perimeter walls |
| 0·71 | | & |
| | | Ddt Bk. wall, thickness: 215, mild stocks, a.b.d. |

Facework on one or both sides of a wall is included in the brick wall item as SMM F10.1.1-3.1.0. Facework is any work in bricks or blocks finished fair (SMM F10.D2). The girth of the perimeter walls has already been calculated in waste. The average height of the facework below damp-proof course could be determined in a variety of ways to give slightly varying answers. The student must however learn not to split hairs, particularly when it involves highly time consuming techniques and to keep a sense of proportion. The approach adopted in this example is to take 6 spot heights, to average them and add 75 mm to take the facings just below ground level. The description includes particulars of the kind and quality of facing bricks, type of bond, and composition, mix and method of pointing (SMM F10.S1-4). The adjustment of the unfaced walls follows.

| | 45·37 | | Dpc, width ≤ 225, hor., single layer of hessian-based bit. felt to BS 743, ref. A, & bedded in c.m. (1:3). (perimeter walls |
| | 0·22 | | |
| 2/ | 8·55 | | |
| | 0·10 | | (divdg. walls |
| 2/ | 0·65 | | |
| | 0·11 | | (projs. on divdg. walls |

The description of the damp-proof course is to include the kind, quality and substance of damp-proof material, number of layers, and composition and mix of bedding material, measured in m², stating the plane and whether ≤ or > 225 mm wide (SMM F30.2.1.3.0).
Follow with the damp-proof courses to the internal walls and projections.

| | | Replacement of topsoil outside bldg. | |
| | | gth. of ext. wall | 45·370 |
| | | add corners to wall 4/215 | 860 |
| | | add corners to fdn. sprd. 4/½/315 | 630 |
| | | | 46·860 |

The topsoil is to be replaced in the void surrounding the building and the girth is measured on its centre line. This item is measured in m³ as filling to excavations in

| | 46·86 | | Fillg. to excavns., av. thickness ≤ 0·25 m, from on site spoil heaps, topsoil. |
| | 0·16 | | |
| | 0·15 | | |

| | | Fillg. under flr. | |
| | | | slabs |
| | | lower end | 850 |
| | | add topsoil | 150 |
| | | | 1·000 |
| | | less conc. 150 | |
| | | h.c. 150 | 300 |
| | | | 700 |
| | | upper end | 400 |
| | | add topsoil | 150 |
| | | | 550 |
| | | less conc. 150 | |
| | | h.c. 150 | 300 |
| | | | 250 |
| | | | 700 |
| | | 2) | 950 |
| | | av. | 475 |

| 3/ | 4·50 | | Fillg. to make up levs., av. thickness > 0·25m obtnd. off site, selected. excvtd. mat., consoldtd. in 150 th. layers. |
| | 8·55 | | |
| | 0·48 | | |

| 3/ | 4·50 | | Compactg. fillg. |
| | 8·55 | | |

### Floor

| 3/ | 4·50 | | Fillg. to make up levs., av. thickness ≤ 0·25 m, obtnd. off site, gravel rejects blinded w. hoggin. |
| | 8·55 | | |
| | 0·15 | | |

| 3/ | 4·50 | | Damp-prfg. membrane, hor., polythene 1000 gauge, ld. on blinded h.c. |
| | 8·55 | | |

accordance with SMM D20.9.1.2.3. If the topsoil was spread on site in layers, a similar approach would be adopted (SMM D20.10.1.2.3).

The depths of fill below the hardcore and concrete slabs are calculated at the lower and upper ends of the building and the average depth computed from them. All filling is measured in m³ classified in the prescribed thickness ranges and giving the other particulars listed in SMM D20.10.2.3.4.
The specification requirement of consolidated layers not exceeding 150 thick is included in the description. No deduction has been made for the volume occupied by the attached piers on the dividing walls, because of the small quantities involved and this will help to compensate for the extra labour in consolidating fill around them.

Compacting filling item as SMM D20.13.2.2.0.

Hardcore filling is measured in m³ giving the particulars prescribed in SMM D20.10.1.3.1. Same approach adopted with regard to deductions for piers as with earth filling. The waterproof membrane is measured in m², stating the pitch (SMM J40.1.1.0.0), with a full description of the materials and nature of the base on which applied (SMM J40.S1-2).

## STEPPED FOUNDATIONS (Contd.)

| | | | |
|---|---|---|---|
| | | 4·500 | The vertical membrane is taken as a linear item to abutments ≤ 200 mm girth (SMM J40.3.2.0.0). |
| | | 8·550 | |
| | | 2/13·050 | |
| | | 26·100 | |
| 3/ | 26·10 | Damp-prfg. membrane to vert. abuts., polythene 1000 gauge, ld. between bwk. & edge of conc. slab, ≤ 200 gth. | It is necessary to link the horizontal damp-proof course in the walls with the horizontal membrane laid over the hardcore. |
| 3/ | 4·50<br>8·55<br>0·15 | In situ conc. bed thickness ≤ 150 (21N/mm² – 20agg.) | Concrete beds are measured in m³, giving the prescribed thickness classification and including laying on earth or hardcore, where unblinded (SMM E10.4.1.0.5).<br>No deduction has been made for the voids occupied by the attached piers, even although the void allowance in SMM E10.M1d (≤ 0·05 m³) does not apply as the voids are at the boundaries of the measured areas (SMM General Rules 3.4), as they are each so small (0·005 m³) and make the laying of the concrete more difficult. |
| 3/ | 4·50<br>8·55 | Trowellg. surf. of conc. | Trowelling the surface of concrete is so described and measured in m² (SMM E41.3.0.0.0). |

1·9

DRG. 2

PILING

### A

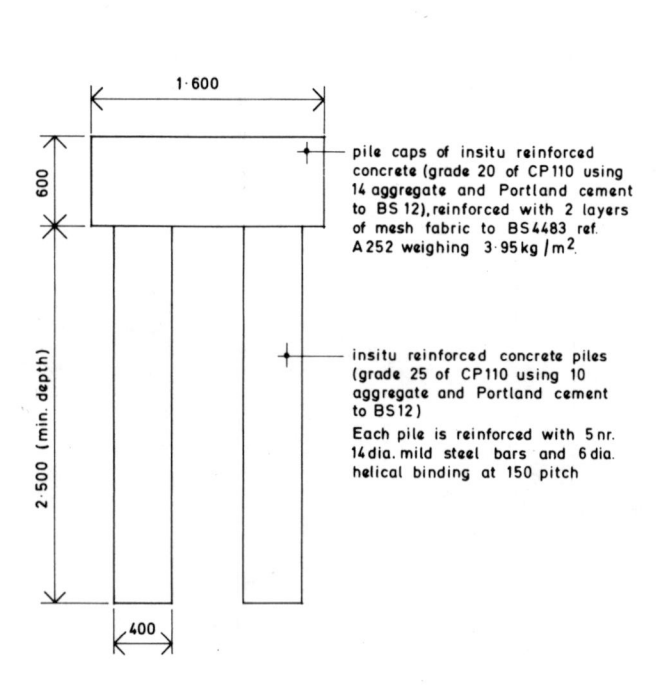

**ELEVATION**

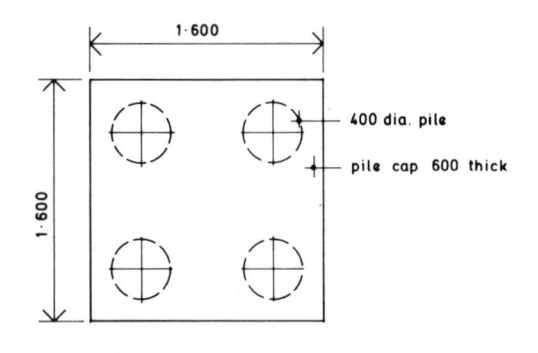

- 1·600
- 600
- 2·500 (min. depth)
- 400

pile caps of insitu reinforced concrete (grade 20 of CP 110 using 14 aggregate and Portland cement to BS 12), reinforced with 2 layers of mesh fabric to BS 4483 ref. A252 weighing 3·95 kg/m².

insitu reinforced concrete piles (grade 25 of CP 110 using 10 aggregate and Portland cement to BS 12) Each pile is reinforced with 5 nr. 14 dia. mild steel bars and 6 dia. helical binding at 150 pitch

**PLAN  Scale  1:50**

- 1·600
- 1·600
- 400 dia. pile
- pile cap 600 thick

16 Nr PILE CAPS EACH SUPPORTED BY 4 Nr. BORED CAST-IN-PLACE CONCRETE PILES (non-percussion method)

### B

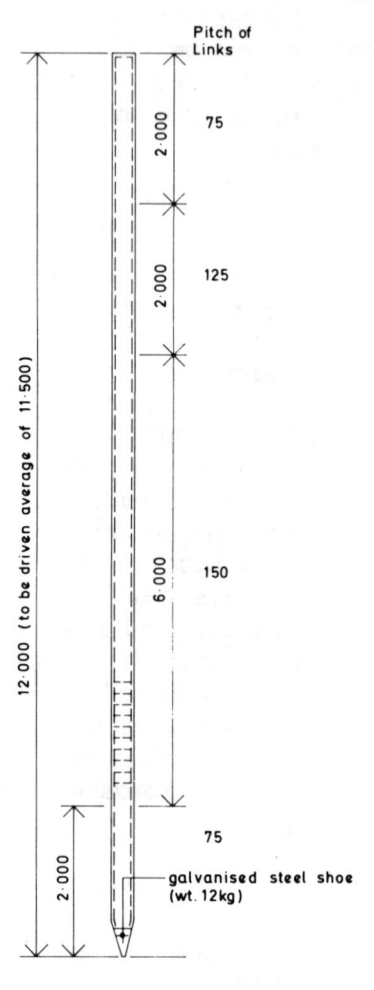

**ELEVATION  Scale  1:100**

Pitch of Links
- 75
- 2·000
- 125
- 2·000
- 150
- 6·000
- 75
- 2·000

12·000 (to be driven average of 11·500)

galvanised steel shoe (wt. 12 kg)

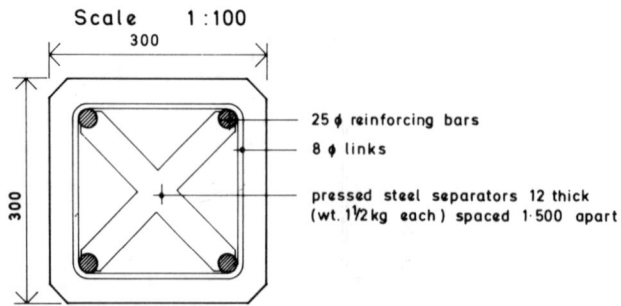

**CROSS SECTION  Scale  1:10**

- 300
- 300
- 25 φ reinforcing bars
- 8 φ links
- pressed steel separators 12 thick (wt. 1½ kg each) spaced 1·500 apart

48 Nr. PRE-FORMED CONCRETE PILES

### C

**PLAN  Scale  1:1000**

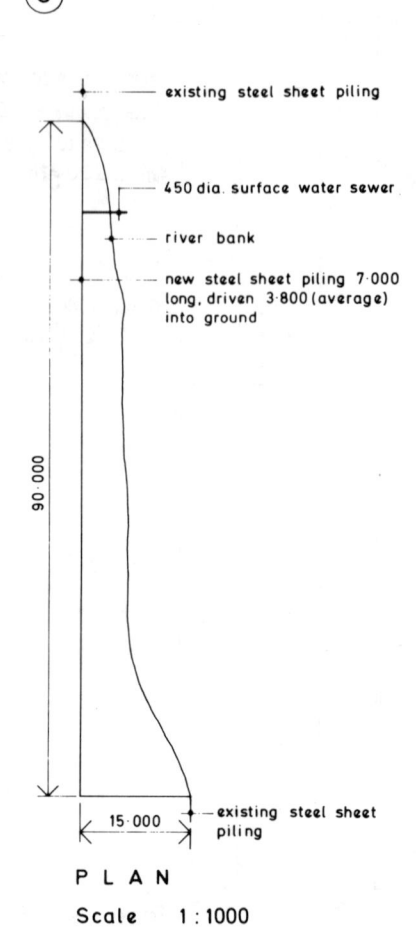

- existing steel sheet piling
- 450 dia. surface water sewer
- river bank
- new steel sheet piling 7·000 long, driven 3·800 (average) into ground
- 90·000
- 15·000
- existing steel sheet piling

INTERLOCKING STEEL PILES

## PILING

(A) **BORED CAST IN PLACE CONCRETE PILES**

The follg. in bored cast in place conc. perm. piling to stan. bases to factory extension., commencg. at 750 mm below e.g.l. (84·250). The subsoil is Keuper Marl & grdwater level at 6.7.89 was 83·500. No ulgrd. or above grd. servs. exist in the area of the pilg. operatns.

_Bored cast in place conc. continuous piles_

400 φ in situ conc. ordinary prescribed mix grade C25P–10 agg. reinfd. w. m.s. bars.

| 16/ | 4 | Total nr. of piles, from 750 below nat. g.l. |
|---|---|---|

```
         2·500
add. connectn. w. pile cap  400
         2·900
```

| 16/4/ | 2·90 | Total concreted len. of piles. & Total len. of piles (max. depth 2·90 m). |
|---|---|---|

| 16/ | 0·50 | Cuttg. off tops of 400 φ piles. (16 nr) |
|---|---|---|

---

In practice location drawings would normally be prepared showing the general piling layout, positions of different types of piles, and work within the site, and adjoining buildings and existing services (SMM D30.P1).

Information provided includes nature of the ground and groundwater level as required by SMM D30.P2, and commencing levels as SMM D30.P3.

Plant items are included in the Preliminaries/General Conditions Bill (SMM A43). A sub-heading stating the nominal diameter and the materials of which the piles are composed will help the contractor and avoid repetition in the item descriptions.

Total number of piles and commencing surface stated as SMM D30.1.1.1.2, and using the same terminology as in the SMM 7 Library of Standard Descriptions.

There are 16 nr pile caps each supported by 4 piles. Total concrete lengths of piles given in metres as SMM D30. 1.1.2.2, while the total length and maximum depth follows in the next item (SMM D30.1.1.3. 2), measured from the commencing surface to the bottom of the bored shaft (SMM D30.M1).

Maximum depth is the depth of the deepest pile in the group.

Cutting off tops of piles is measured in metres, stating the number of piles (SMM D30.7. 1.1.0). Cutting off tops of piles is deemed to include preparation and integration of reinforcement into pile cap and disposal (SMM D30.C5).

2·1

---

| 16/ | 4 |  |
|---|---|---|

| 16/4/3/ 22/7 | 0·50 0·50 1·00 | |
| 16/4/ 22/7 | 0·20 0·20 1·50 | |

```
                    2·500
add unto pile cap    300
bend                 300
                    3·100
```

| 16/4/5/ | 3·10 | |

```
                400 φ
less cover 2/40  80
                320 φ
               circumf.
320 × 22/7  =  1·006
add o/laps, say  60
                1·066

        150)2·500
        17+1 = 18
```

| 16/4/18/ | 1·07 | |

| 16/ | 1·60 1·60 0·60 | |

---

E.O. piles for enlarging base, 1·00 m extreme diam.

[diagram: trapezoidal enlarged base, 400 top, 1·000 bottom, 1·000 height]

_enlarged base to each pile_

```
                    2·500
less enlarged base  1·000
                    1·500
```

Disposal of surplus excvtd. mat. on site in perm. spoil heaps av. 100 mm from pilg.  (enlarged bases)

(upper secs. of piles)

_Reinft._

Ms bar reinft. 14 φ nom. size, strt. & bend, to BS 4449 in piles.

Ms helical reinft. bars, 6 φ nom. size, to BS 4449; in 400 φ piles.

_Pile caps_

Exc. for pile caps; max. depth ≤ 1·00 m. &

Disposal of excvtd. mat. off site.

2·2

---

## EXAMPLE 2

Enumerated extra over piles for enlarged bases as SMM D30.5.2.1.0, stating the extreme maximum diameter, to cover the additional excavation and concrete required.

Disposal of excavated material on site in permanent spoil heaps or spread on site is given in m³, describing the location of the deposits or the average distance from the excavation in m or km (SMM D30.9.1.2.1).

Calculate total length of steel bars in waste, allowing for 300 mm penetration into pile cap and a bend to provide increased key.

Note the order of timesing – number of pile caps; piles to each cap; and bars in each pile. The reinforcement particulars must include the kind and quality of steel (SMM D30.S7) and the nominal size (SMM D30.8.1.0.0).

Calculation of circumference of helical reinforcement and of number of bars at the specified pitch of 150 mm.

Helical reinforcement bars are separately categorised as SMM D30.8.2.1.0, stating the diameter of the piles.

Excavation for pile caps is given in m³ (SMM D20.2.7.2.0).

Excavated material disposal is measured in accordance with SMM D20.8.3.1.0.

| | | | | | | | |
|---|---|---|---|---|---|---|---|
| | PILING (Contd.) | | | 16/2/ | 1·50 | Stl. fabric reinft. to BS 4483 ref. A252, weighing 3·95 kg/m², w. min. 150 laps. | Steel fabric reinforcement is measured in m² stating the mesh reference and weight per m² and minimum laps as SMM E30.4.1.0.0 and E30.S4. Fabric reinforcement is deemed to include laps, tying wire, all cutting and bending, and spacers and chairs which are at the discretion of the contractor (SMM E30.C2). |
| | BORED CAST IN PLACE CONCRETE PILES | | | | 1·50 | | |
| 16/ | 1·60 1·60 | Compactg. botts. of excavns. | Surface treatment of bottoms of excavations is measured in m² and classified as SMM D20.13.2.3.0. | | | | |
| 16/4/ | 1·60 0·60 | Earthwk. suppt. max. depth ≤ 1·00 m; dist. between opposg. faces ≤ 2·00 m. | Earthwork support is measured in m² stating the maximum depth range and the distance between opposing faces as SMM D20.7.1.1.0. No working space allowance is measured as the face of the excavation is less than 600 mm from the face of the formwork (SMM D20.M7). | 16/ | 1·60 1·60 | Trowellg. surf. of conc. | Treating the surface of the concrete in m² as SMM E41.3.0.0.0. |
| | | | | | 20 h | Authorised delays, rig standing time. | Delays are only measured where specifically authorised (SMM D30.M7) and are deemed to include associated labour (SMM D30.C7). |
| 16/ | 1·60 1·60 0·60 | In situ conc. isoltd. fdns. ord. prescribed mix grade C20P-14agg. reinfd. | The particulars of the concrete are detailed as SMM E10.S1 and reinforced work is so described. Pile caps would seem to be covered by the classification of isolated foundations as SMM E10.3.0.0.1. Alternatively the concrete might be described as poured on or against earth and the following item of formwork omitted. | | Item | Test 400 ⌀ pile by kentledges w. total load on ea. pile of 42t maintained for 24 hrs. | The testing item often includes the method, load and rate of loading. SMM D30.11.1.0.0 requires details of pile tests to be stated. For example a 1·00 m diameter pile with a design load of 100 tonnes could have the load applied in 10 tonne increments at hourly intervals and for the load to be removed after 24 hours, in equal increments over a 6 hour period. |
| 16/4/ | 1·60 | Fwk. to sides of fdns., plain vert., ht. 500 mm – 1·00 m. | When measuring formwork, the term foundations includes pile caps (SMM E20.D4). It is classified as SMM E20.1.1.4.0. | | | | |
| | less cover 2/50 | 1·600 100 1·500 | Allowance has been made for concrete cover when measuring the fabric, but this might be omitted in practice because of the small quantities involved. Area is twiced as there are two layers. | | | | |

2.3   2.4

## PILING (Contd.)

### B. PREFORMED CONCRETE PILES

The follg. in 48 nr. preformed precast concrete perm. piles in prepn. for new office bldg.
The gen. pilg. layout & posns. of the piles, wk. within the site & xtg. services are shown on Drg. HB24A.
The grd. generally consists of topsoil 100 mm th. & soft clay 2 m dp. overlaying sand and gravel on a reasonable lev. site.
Details of boreholes are shown on Drg. HB24B.
Wk. begins from e.g.l. as shown on Drg. HB24B.
Groundwater level was established on 9.6.89 as 4.60 m below e.g.l. (124.80).
A 375 ⌀ sewer crosses the site in the position shown on Drg. HB24A.

General information covering the piling and ground conditions are given to meet the requirements of SMM D31.P1-3. An alternative approach has been adopted to that used for the bored cast in place concrete piles.
Note that particulars are required of groundwater level and the date when it was established (pre-contract water level). This will be re-established when work is commenced (post contract water level).
In the absence of trial pits or boreholes, which are normally shown on location drawings, an assumed description of the ground and strata will be provided as a basis for pricing by the contractor (SMM D.20.P1).
Plant items are included in the Preliminaries/General Conditions Bill (SMM A43).

_Preformed reinfd. conc. piles, 300 x 300, w. chfd. edges & a min. comp. stress of 50N/mm² at 28dys., reinfd. as Dwg. B & w. g.s. pile shoes ea. 600 lg. & weighing 12 kg._

| | | |
|---|---|---|
| 48/ | 1 | Total nr of piles driven, specfd. len. 12.00 m; from nat. g.l. |

Enumerated item giving the details prescribed in SMM D31.1.1.1.0.

| | | |
|---|---|---|
| 48/ | 11.50 | Total driven len. of piles. |

The total driven length is given in m (SMM D31.1.1.2.0). The tops of the piles are to project 500 mm above natural ground level.

| | | |
|---|---|---|
| 48/ | 0.30 | Cuttg. of tops of piles, total len. (48 nr) |

Linear item as SMM D31.10.0.1.0, stating the number of piles.

It is deemed to include the preparation and integration of reinforcement into pile cap or ground beam and disposal (SMM D31.C4).
Driving heads and shoes are deemed to be included (SMM D31.C1), but details are required as SMM D31.S3.

| | | |
|---|---|---|
| 48/ | 0.30 | Disposal of surplus excvtd. mat. off site. |
| | 0.30 | |
| | 11.50 | |

Disposal of excavated material, measured as SMM D31.11.1.1.0. The volume is calculated from the nominal cross-sectional size of piles and driven depth (SMM D31.M5), which seems rather illogical as stated in the text.

| | | |
|---|---|---|
| 20 | h | Authorised delays, rig standing time. |

Where delays are anticipated provision should be made for standing time for the driving rig to secure an hourly rate (SMM D31.12.1.0.0).

Delays are only measured where they are specifically authorised (SMM D31.M6), and are deemed to include associated labour (SMM D31.C5).

| | |
|---|---|
| 48 | Test 300 x 300 piles by kentledges w. total load on ea. pr. of piles of 100t applied in 10t increments at hourly intervals, maintg. ld. for 24 hrs & removg. in equal increments over 6 hrs. |

Testing item giving full details as provided for in SMM D31.13.1.0.0.

2.5      2.6

| | | | | | | | | |
|---|---|---|---|---|---|---|---|---|
| | Ⓒ | | INTERLOCKING STEEL PILES | | | | | |

© INTERLOCKING STEEL PILES

The follg. in perm. interlockg. stl. piles driven vertically into the river bk. in the positions shown on location drg. X118B, w. the upper surf. at e.g.l. 900 mm above HWOST and 4·20 m above LWOST.

*The position of steel sheet piling will normally be obtained from location drawings. Where work is to be carried out near canals, rivers or tidal waters, the level of the ground in relation to the normal levels of the canal or river or to the mean Spring levels of high and low tidal water shall be stated, and flood levels where applicable. Also the levels from which the work is to begin and from which measurements have been taken (SMM D32.P1-3).*

*Plant items are included in the Preliminaries/General Conditions Bill (SMM A43).*

Interlockg. stl. piles to BS4360, grade 50B, type 3, weighg. 155 kg/m²

         90·000
         15·000
        105·000

*The description of interlocking piles should state the section modulus and cross-sectional size or section reference (SMM D32.2.1).*

*Calculate total length of piling in waste.*

| 105·00 | | |
| 7·00 | Total area of piles specfd. len. ≤ 14·00 m. | *Total area of piles is given in m², and classified in one of three length stages as SMM D32.2.1.1.0.* |

| 105·00 | | |
| 3·80 | Total driven area of piles. | *Total driven area of piling is measured in m² as SMM D32.2.1.4.0.* |

| 2/ 7·00 | E.o. interlockg. stl. piles for corner pile. | *Linear extra over items for corner piles, as SMM D32.3.1.1.0, one at right angled turn and the other at the right angled connection with the existing piling.* |

| 7·00 | E.o. interlockg. stl. piles for junctn. pile. | *Linear extra over item for junction pile as straight connection of new to existing piling as SMM D32.3.2.1.0* |

2·7

| 1 | | Cut interlockg. stl. piles to form hole 525⌀. | *Enumerated item for cutting hole in piles for the sewer outfall as SMM D32.7.1.0.0, stating the size of the hole. It is necessary to allow for the thickness of the pipe when determining the hole size.* |

| 15 | h | Authorised delays, rig standing time, interlocking piles. | *Authorised delays given in hours in accordance with SMM D32.8.1.2.0; wording of item taken from SMM7 Library of Standard Descriptions.* |

| 105 | | Test interlockg. stl. piles by a vertical load totallg. 60t applied to 1·000 m len. of piles in 10t increments at hourly intervals, maintaing. ld. for 24 hrs. & removg. ld. in equal increments over 6 hours. | *Testing item as SMM D32.9.1.0.0, stating details of method to be used.* |

2·8

## DRG. 3

# UNDERPINNING

**Notes:**

Trial holes indicate no rock in vicinity of extension

Standing water encountered at 2·500 m below ground level adjoining existing building on 15 June 1981

Existing ground level is 150 below existing ground floor level

Level site

Backfilling to be in selected excavated material except behind new wall in underpinning which is to be in concrete (1:10)

Underpinning to be executed in equal stages not exceeding 1·250 long

Topsoil is not required to be preserved

- 0 / 150 / 50 — turf / topsoil
- 1·110 — loam containing loose small stones
- 2·500 — groundwater level
- 3·000 — light blue clay

TRIAL HOLE 3·000 DEEP

### PLAN

- extent of underpinning
- 102·5 / 215
- one brick inner skin, 20 thick 3 coat asphalt tanking and half brick outer skin
- NEW BASEMENT / EXISTING BASEMENT
- ramp down
- 1·350
- 5·450
- 338 / 20
- line of concrete extent of underpinning

### SECTION

PROPOSED EXTENSION TO CONCERT ROOM | EXISTING CONCERT ROOM

- 50 floor finish and hollow beam floor (150 thick)
- line of existing building
- existing ground floor
- new lintel over opening
- existing one brick wall
- render existing brickwork in 2 coat cement and sand (1:3) plain face
- brick on edge coping
- 2·100
- half brick wall 500 high
- cut off projecting foundation
- new ramp
- pin up in cement mortar (1:3)
- concrete backfill (1:10)
- one brick wall in class B engineering bricks
- new underpinning to existing structure
- concrete grade C25P-20 aggregate
- 800 / 300 / 700

New floor construction paving 50 thick laid monolithically with 100 reinforced concrete bed on 30 thick 3 coat asphalt on 150 reinforced concrete sub bed on 150 layer of hardcore blinded with fine material

Scale 1:50

## UNDERPINNING

The follg. in underpinning. 215 ex. bk. wall 5·45 m lg. adjg. new extnsn. to a depth of 2·48 m below e.g.l. & 1·10 m below base of xtg. fdns, as shown on Drg. 3. The wk. to be exectd. in lens. n.e. 1·00 m from outside the xtg. bldg. in alt. secs. w. not more than 3 secs. exposed at any one time. The subsoil is lt. blue clay & grdwater lev. was 2·500 below g.l. on 15.7.89. No u/grd. or above grd. servs. exist in the area of bldg. operations.

*Underpinning work forms a separate section in the bill of quantities. Locational drawings should show the location and extent of the work, details of the existing structure to be underpinned and ground particulars. The limit of length carried out in one operation and the number of sections the contractor is permitted to undertake at one time (SMM D50. P1-3).*

Excavn.
| | | len. |
|---|---|---|
| add | | 5·450 |
| inner leaf | 215 | |
| asp. | 20 | |
| outer leaf | 102·5 | |
| fdn. proj. | 200 | |
| | 2) 537·5 | 1·075 |
| | | 6·525 |
| | 6) 6·525 | |
| u/pping. lens. | | 1·088 |

*Calculation of length of underpinning work to existing structure, and its division into appropriate working lengths.*

Item — Tempy. suppt. to xtg. 215 bk. wall to be u/pinnd. for a len. of 6·53 m & m/gd. to xtg. structure as nec.

*Temporary support for existing structure as SMM D50.1.1.0.0. In this case the form of the temporary work is made the responsibility of the contractor and includes any necessary making good as SMM D50.S1.*

Prelimy. trench.
| | | depth |
|---|---|---|
| scrd. & flr. slab | 200 | |
| bast. flr. to clg. | 2·100 | |
| bast. flr. pavg. | 50 | |
| r.c. bed | 100 | |
| asp. | 30 | |
| r.c. sub-bed | 150 | 330 |
| | | 2·630 |
| less grd. lev. to grd. flr. lev. | | 150 |
| | | 2·480 |
| Additional depth to base of u/p. bwk. | 800 | |
| conc. fdn. | 300 | 1·100 |
| | | 3·580 |

| | |
|---|---|
| | 700 |
| | 215 |
| | 2) 485 |
| | 243 |

width allowance 2·000
(total depth > 3 m)
add proj. of retained fdn. 243
2·243

*Calculate depth of preliminary trench from figured dimensions showing all relevant calculations in waste.*

3.1

## EXAMPLE 3

| 6·53 | Exc. prelim. tr. max. depth ≤ 4·00 m, from o.s. only. |
| 2·24 | |
| 2·48 | |

&

Disposal of excavtd. mat. off site.

*Excavation to preliminary trenches extending down to the base of the existing foundation is classified as SMM D50.2.1.4.2. and D50.D1 with the maximum depth listed in the appropriate depth range and stating whether excavation is from one or both sides. The width allowance of 2·00 m results from the application of SMM D50.M1, where the total depth of excavation (top of preliminary trench to base of underpinning work) > 3·00 m, to which is added the projection of the retained foundation.*

*The disposal of the excavated material from the site is a separate cubic item as SMM D20.8.3.1.0.*

| 6·53 | Earthwk. suppt. to faces of prelim. tr. max. depth ≤ 4·00 m & dist. between opposing faces 2·00 - 4·00 m. |
| 2·48 | |
| 2/ 2·24 | |
| 2·48 | (ends of tr.) |

*Earthwork support to preliminary trenches is described separately (SMM D50.4.1.0.0), with maximum depth classification and the distance between opposing faces in accordance with SMM D20.7.3.2.0. Only one face is measured as the other consists of the existing brick wall.*

| 6·53 | Cuttg. away xtg. proj. fdns., masonry, max. 150 x 225 dp. |

&

Ditto. conc., max. 243 x 300 dp.

*Linear item for cutting away existing projecting foundations, giving the maximum width and depth of projections for both brickwork and concrete (SMM D50.5.1 - 2.1.0).*

Underpinning pits
width allce. 2·000
add width of retained fdn. 700
2·700

*Excavating underpinning pits is given as a separate item in accordance with SMM D50.2.2.3.2.*

| 6·53 | Exc. u/p. pits, max. depth ≤ 2·00 m, from o.s. only. |
| 2·70 | |
| 1·10 | |

&

Fillg. to excavns. av. thickness > 0·25 m, arisg. from excavns.

*The disposal of the excavated material from this trench is taken as filling in the first instance, but parts of it will be adjusted subsequently, when measuring the concrete and brickwork.*

3.2

## UNDERPINNING (Contd.)

| | | | | | |
|---|---|---|---|---|---|
| 7/ | 6.53<br>2.70<br>1.10 | E.o. excavn. irrespective of depth for excavtg. below grdwater. lev. | Excavating below groundwater level is given in m³ as extra over all types of excavation irrespective of depth (SMM D20.3.1.0.0). | | |

7/ 6.53
1.10
0.70
0.10

Earthwk. suppt. to faces of u/p pits, max. depth ≤ 2.00 m & dist. between opposg. faces: 2.00 – 4.00m, below gwl.   (ends

Earthwork support classified as SMM D50.4.2.0.0, and using same terminology as SMM7 Library of Standard Descriptions. Although the Library descriptions are sometimes at variance with the wording of SMM7. It is measured to the back, front and both ends of the underpinning pits and also between each section of the underpinning (SMM D50.M6).

6.53
1.10

Ditto., left in.

Earthwork support left in shall be so described (SMM D20.7.2.2.6).

6.53
0.70

Compactg. bott. of excavn.

Surface treatment of bottom of excavation to receive concrete is measured in m² (SMM D20.13.2.3.0).

6.53
0.70
0.30

In situ conc. fdns., ord. prescrbd. mix: grade C25P-20 agg. poured on or against earth.
&
Ddt. Fillg. to excavns. a.b.d.
&
Add Disposal of excvtd. mat. off site.

The particulars of in situ concrete shall satisfy the requirements of SMM E10.S1 and in this case an ordinary prescribed mix from CP 110 has been selected. Note the requirements to include 'poured on or against earth' (SMM E10.1.0.0.5).
The concrete is followed by the adjustment of soil disposal which was originally measured as filling.

6.53

Fwk. to sides of fdns., plain vert., ht. 250 – 500.

        ht.
        1.100
less fdn. 300
        800

Formwork to sides of foundations is so described in accordance with SMM E20.1.1.3.0 and where less than 1.00 m high, is measured in metres, giving the appropriate height classification.

6.53
0.80

Bk. wall, thickness: 215, class B eng. bks to BS 3921 in Eng. bond in c.m. (1:3), used as fwk. & inc. any nec. tempy. struttg.

Brickwork is given in m² stating the thickness and classified as SMM F10.1.1.1.3.

3.3

---

In this case the brickwork is to serve as formwork to the concrete backing. The brickwork is deemed vertical unless otherwise described (SMM F10.D3). Particulars are to include the kind, quality and size of bricks, type of bond and composition and mix of mortar (SMM F10.S1-3). The standard size of bricks is 215 x 102.5 x 65 mm and this is usually given in the preamble. Follow with soil disposal adjustment to avoid it being omitted.

6.53
0.22
0.80

Ddt. Fillg. to excavns. a.b.d.
&
Add Disposal of excvtd. mat. off site.

6.53
0.24
0.80

In situ conc. (1:10) as backg. to bk. wall, thickness: 150 – 300, poured on or against earth.
&
Ddt. Fillg. to excavns. a.b.d.
&
Add Disposal of excvtd. mat. off site.

There is no concrete category in SMM E10 which specifically covers this class of work. A suitable description is framed, incorporating the appropriate thickness classification, operating the provision in SMM General Rules 1.1.
Adjustment of soil disposal follows for the volume occupied by the concrete.

2/ 0.80

Fwk. to ends of backg., plain vert., width ≤ 250.

        700
        243
        457

Formwork to ends of concrete backing to brick wall taken as a linear item, adopting the same height ranges as for sides of upstands, but expressed as a width range, using the discretion given in SMM General Rules 1.1.

6.53

Prep. u/s of the xtg. wk., 457 wide, to rec. pinng. up of the new wk.
&
Wedgg. & pinng. new one bk. wall to u/s of xtg. conc. fdns., in stiff c.m. (1:3) 25 th.

Linear item measured in accordance with SMM D50.6.1.0.0.

Linear item as SMM F30.7.1.0.0, stating the thickness of the work and describing the materials to be used for wedging and pinning.

Item   Disposal of surf. water.   See SMM D20.8.1.0.0.

Item   Disposal of grdwater.

This additional item is required as this section of the excavations is below groundwater level (SMM D20.8.2.0.0).

3.4

# 3 BRICKWORK, BLOCKWORK AND MASONRY

## BRICKWORK AND BLOCKWORK

*General requirements*

It is necessary to provide location drawings or further drawings accompanying the bill of quantities incorporating plans of each floor level and principal sections showing the provision of and materials used in the walls, together with external elevations showing the materials used (SMM F10.P1).

*General brickwork and blockwork*

An adequate description of the bricks or blocks, incorporating the kind, quality and size, should be given in a preamble, bill sub-heading or measured item description as outlined in SMM F10.S1, sufficient to enable the contractor to determine the degree of hardness of bricks or blocks and hence the relative difficulty in cutting them, now that all cutting is deemed to be included in the brickwork and blockwork rates (SMM F10.C1b).

Different thicknesses of brickwork or blockwork are kept separate and are each measured in $m^2$. Four separate classifications are provided in SMM F10.1–4, namely walls, isolated piers, isolated casings, and chimney stacks. In addition, work built against or bonded to other work, used as formwork or built overhand, shall be kept separate because of the extra cost involved (SMM F10.1.1.1.1–4). There is also separate provision for battering walls and those tapering on one or both sides (SMM F10.1.1.2–4.0), and forming and closing cavities in hollow walls (SMM F30.1.1.1.0 and F10.12.1.1.0). All work is deemed vertical unless otherwise described (SMM F10.D3).

It will be noted that projections are measured in metres stating the width and depth of projection and the plane (SMM F10.5.1.1–3.0). Projections consist of attached piers (whose length on plan is ≤ four times their thickness), plinths, oversailing courses and the like (SMM F10.D9). For example, a projecting pier 675 × 113 mm in common bricks would be measured in metres as 'Brick projection, width: 675, depth of projection: 113, vertical, commons in English bond in cement mortar (1:3)'.

*Brick and block facework*

Brick and block facework is included in the walling item, stating whether it is on one or both sides and giving the thickness of the wall (SMM F10.1.2–3.1.0). Full particulars of the bricks or blocks, bond, mortar and pointing must be given (SMM F10.S1–4). Facework is any work in bricks or blocks finished fair (SMM F10.D2).

Descriptions of facing and other selected bricks often include the basic price or net price per thousand, delivered to the site. If the term 'prime cost' is used this is deemed to be exclusive of any profit required by the general contractor and provision must be made for its addition (SMM A.52). An alternative approach is to keep the supply of bricks separate from the laying item using the method that follows.

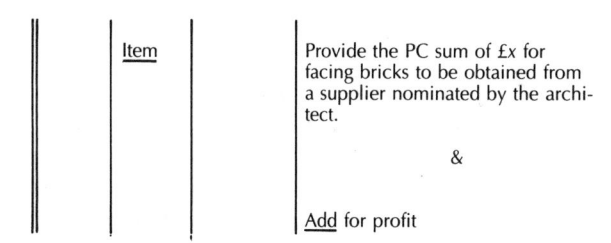

| | Item | Provide the PC sum of £x for facing bricks to be obtained from a supplier nominated by the architect. |
| | | & |
| | | Add for profit |

The brickwork item that follows indicates that the bricks are supplied by a nominated supplier: 'Brick wall, facework one side, thickness: 102.5, LBC facing bricks (from nominated supplier)..........'

Facework ornamental bands and the like are measured in metres, stating the type and whether flush, sunk or projecting, including depth of set back or set forward, plane and width and whether extra over the work in which they occur (SMM F10.13.1–3.3.1), while facework quoin descriptions include the mean girth (SMM F10.14.1–3.1.0). Facework ornamental bands include brick-on-edge bands, brick-on-end bands, basket pattern bands, moulded or splayed plinth cappings, moulded string courses, moulded cornices and the like (SMM F10.D11). Labours in returns, ends and angles are deemed to be included (SMM F10.C1f).

There are also rules for the measurement of arches (SMM F10.6), and facework sills, thresholds, copings and steps (SMM F10.15–18), all of which are reasonably self-explanatory.

*Other classes of work*

Work in boiler seatings and flue linings are measured in m² stating the thickness (SMM F10.8 & 9). Isolated chimney shafts are measured separately under a heading, stating the number, size on plan, shape and overall height, as SMM F10.7.1.1.1, and this approach has been adopted in example 4, even although the chimney shaft is not isolated from the building, as it cannot realistically be classed as a chimney stack. This example also illustrates the application of many of the brickwork rules previously described.

Other brickwork examples are given in *Building Quantities Explained* covering a basement, hollow walls, curved screen wall and a chimney breast, stack and fireplace. Taken together these give an adequate coverage of this class of work. A good sequence of measurement is vital, usually following the order of construction on site, and it is good practice to mark each section on the drawings as it is taken off, often using coloured inks, and to prepare a rough take off list prior to measuring. When measuring an irregular-shaped building, check to see that the sum of all the individual portions equates to the total length.

## DAMP-PROOF COURSES AND SUNDRIES TO BRICKWORK AND BLOCKWORK

Descriptions of damp-proof courses must include all the particulars listed in SMM F30.S4–6; a typical description is given in example 4. Work ≤ and > 225 mm wide are kept separate, but both categories are measured in m², stating the plane and whether in cavity trays (SMM F30.2.1–2.3.1). All cutting and forming grooves, throats, rebates and holes are deemed to be included in the brickwork and blockwork rates (SMM F10.C1b–c), hence no additional items are required for building in or cutting and pinning ends of lintels, timbers, steel sections and the like.

Enumerated items are however required for cutting or forming holes for ducts or pipes for service installations, stating the nature and thickness of the structure, the appropriate girth or nominal size range, shape and making good where necessary (SMM P31.20.1–2.1–3.1–4). Linear items are included for joint reinforcement (SMM F30.3.1.0.0), weather and angle fillets (SMM F30. 4 & 5.1.0.0), pointing in flashings (SMM F30.6.0.0.0), wedging and pinning new work to old (SMM F30.7.1.0.0), slate and tile creasings and sills (SMM F30.9–10.1.1.0) and flue linings (SMM F30.11.1.0.0). While air bricks, ventilating gratings, soot doors, gas flue blocks and proprietary items are enumerated (SMM F30.12–16).

## MASONRY

*Measurement requirements*

Stone walls and chimney stacks are each measured separately in m², giving the thickness and adopting the appropriate classifications given in SMM F21.1–2.1.1–4.1–18, while isolated and attached columns are each measured in metres giving a dimensioned description (SMM F21.3–4.1.1–4.0). Rough or fair raking or circular cutting are each measured separately in metres, stating the thickness as SMM F21.26–27.1.0.0. Grooves, throats, flutes and rebates on superficial items of masonry and attached piers are taken as linear items stating the size (SMM F21.28–31.1.0.0 and F21.M10), while chases are subdivided between rough and fair categories and given in 150 mm girth stages in metres as SMM F21.32.1–2.1–2.0. It will be noted that large stones or blocks, > 1.50 m long or > 0.50 m³ are so classified.

There is a wide range of linear items, including dimensioned descriptions, embracing:

(1) Quoin stones and jamb stones, categorised into attached, attached with different finish, and isolated (being attached to another form of construction ) (SMM F21.10–11.1–3.1.0 and F21.D13).

(2) Slab architraves and slab surrounds to openings, being slabs which are not bonded to their surrounding work (SMM F21.12–13.1.0.0 and F21.D14).

(3) Bands, corbel courses, copings, handrails, cappings, kerbs and cover stones, stating the plane and including mouldings in the descriptions, although plain bands > 300 mm wide are measured as walling or facework (SMM F21.14–20.1.1–4.0, F21.M7 and F21.D16).

(4) Steps, giving the number and distinguishing between plain and spandrel steps; the spandrel steps having sloping soffits (SMM F21.21.1.1–2.0 and F21.D17).

(5) Arches, giving the number and height of face, width of soffit and shape of arch, and measured the mean girth or length on face, or alternatively they can be enumerated (SMM F21.24.1.0.0 and F21.M8 & 9).

(6) Closing cavities, stating the width of cavity and method of closing and plane (SMM F21.25.1.1–3.0).

Enumerated items include winders and landings (SMM F21.22–23.0.1.0), special purpose stones giving the function and a dimensioned description, with the description stating the smallest block from which each item can be obtained, including one mortar bed and one mortar joint, and in the case of natural stone having regard to the plane in which the stone is to be laid with relation to its quarry bed (SMM F21.33.1.1–2.0 and F21.M11), carvings and sculptures stating the character of work and supplying a component drawing (SMM F21.34–35.1.1.1), and centering giving the appropriate classifications incorporated in SMM F21.36.1–4.1.1–4, and the dimensioned description shall give the shape and width of the surface to be supported, the span of the soffit and, in the case of arches, whether segmental, semicircular, invert and the like, stating the rise (SMM F21.M12).

*Description of masonry*
Each type of stone is kept separate and the billed items are normally preceded by a heading giving the kind and quality of stone, any requirements as to quarry, texture and finish to exposed faces, details of composition and mix of mortar, type of pointing and method of jointing together and fixing, and any requirements as to coatings to backs of stones, stones not set on their natural bed, and type and positioning of metal cramps, slates, dowels, metal dowels, lead plugs and the like (SMM F21.S1–10). The latter fixing items are not measured separately as they are deemed to be included in the stone walling (SMM F21.C1e).

The billed descriptions of stone walling and dressings will incorporate the function or purpose of the stone followed by a dimensioned description. Components serving the same function but having different dimensions or labours will be kept separate.

*SMM7 Code of Procedure for Measurement of Building Works* describes how special purpose stones, covered by SMM F21.33.1.1–2.0, may include purpose made blocks within other measured items, such as purpose made corner blocks to sills, involving the use of stones larger than the sills. They also embrace a variety of blocks which by their very nature are purpose made, such as ornaments; small panels ($\leq 0.1$ m$^2$); caps and bases to columns; kneeler blocks, apex blocks and the like to copings and ornamental band courses; plinth blocks, angle blocks and the like to slab architraves and surrounds to openings; springers, voussoirs and keystones of arches; pier caps and chimney caps; finials; stones forming tracery; stones forming balustrade panels; balusters, half balusters, newels and newel caps; templates, bases, thresholds and hearths; and stones forming shelves, divisions, table tops and work tops.

Dimensioned diagrams showing the shape and dimensions of the work covered by an item may be used in place of a dimensioned description and three stonework examples are illustrated in figure 3 of the *Code of Procedure* (see SMM7 General Rules 5.3).

Labours on masonry fall into two main categories.

(1) texture of the surface, such as sawn, rubbed, tooled vermiculated and polished;

(2) shape of the stone, in which connection the following definitions may prove helpful.

*Plain.* Worked face on the surface of the stone but involving no sinkings.

*Sunk.* Plain face worked below the surface of the stone such as chamfers and rebates (stating number of times it is sunk).

*Moulded.* Sunk face with the surface made up of a combination of curves or fillets.

*Circular.* Curved convex surface.

*Circular sunk.* Curved concave surface.

*Circular-circular.* Spherical convex surface such as a ball finial.

*Circular–circular sunk.* Spherical concave surface such as the soffit of a dome.

*Sunk jointed.* Sunk face that is not exposed and adjoins brickwork or masonry.

*Circular jointed.* As sunk jointed but with curved convex surface.

*Circular sunk jointed.* As sunk jointed but with curved concave surface.

The use of most of these terms is illustrated in example 5A.

*Measurement procedures*
Before starting to measure masonry it is advisable to reference each course alphabetically, working from the bottom upwards. Where a large quantity of masonry is involved, it is better to number each stone or to letter each course and number each stone on each course. Next, insert the height of each course on the drawing, checking the total of the individual heights against the overall height figures on the drawing. It is also good practice to show the bed widths on the drawing.

Once these preparatory steps have been taken, the taking off can proceed, starting with the lowest course. The total of the individual stone lengths on each course should be checked against the overall length and each stone suitably marked on the elevation as it is taken off. It is frequently possible to group a number of different courses together, where the stones are of similar dimensions and finish, but care should be taken to check the individual course heights.

Comparatively thin, rectangular blocks of dressed stone with fine joints, used as facework to brickwork, are termed 'ashlar'. The normal masonry rules of measurement apply, that is, in m² stating the thickness and whether built against or bonded to the brickwork.

Cast stone walling and dressings are measured in a similar manner to natural stone (SMM F22). A typical description is given in example 5C, from which it will be noted that the backing on a reinforced concrete wall to receive the stone slabs is a separate superficial item (SMM M20.1.1.1.0). The slabbing description includes the type of stone, average height of course, type of finish and method of fixing and jointing (SMM F22.S1–10).

Further stone components in a balustrade are measured in example 5B giving examples of linear and enumerated items with their accompanying labours. Fixing dowels and forming mortices in the adjoining stones are deemed to be included (SMM F21.C1c and e).

## RUBBLE WALLING

Rubble walling, consisting of natural stones either irregular in shape or roughly dressed and laid dry or in mortar with comparatively thick joints, is measured in m², in the same classifications as for natural stone walling and dressings (SMM F20.1–2.1.1–4.0).

Note the various particulars that are to be given in the description of rubble walling — kind and quality of stone, such as limestone whitbed, finish, such as rough dressed, type of walling, that is, whether of random or squared stones, and, where coursed, the average height of the courses or maximum and minimum heights of diminishing courses, composition and mix of mortar, type of pointing and method of jointing (SMM F20.S1–10). Facework is included in the walling, stating whether to one or both sides. Rubble facework where applied to a backing material other than rubble walling includes the bonding to other work in the description.

Levelling uncoursed rubble work for damp-proof courses, sills, copings and the like, labours in returns, ends and angles, dressed margins in rubble work, and rough and fair square cutting are deemed to be included without the need for specific mention (SMM F20.C1g, h and j). Rough or fair raking or circular cutting are each measured separately in metres stating the thickness (SMM F20.26–27.1.0.0).

Copings, often formed of rough stones, are measured in metres, giving a dimensioned description and distinguishing between horizontal, vertical and raking copings (SMM F20.16.0.1–4.0), and curved work is kept separate, stating the radii (SMM F20.M4). Arches are measured in metres (stating the number) or enumerated giving the height of face, width of soffit and shape of arch (SMM F20.24.1.0.0).

A typical rubble walling description might read 'Stone wall, thickness: 300, random rubble, uncoursed in picked local sandstone from 'X' quarry, rough dressed and jointed in lime mortar (1:3), and pointed both sides in gauged mortar (1:1:6) with a flush joint'. A short example of the measurement of a random rubble boundary wall is contained in *Building Quantities Explained*.

FOR READER'S NOTES

# DRG. 4

# BRICK CHIMNEY SHAFT

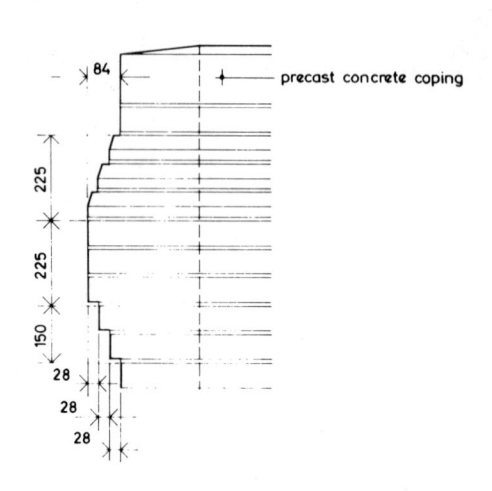

**Ornamental Band Detail B**
Scale 1:20

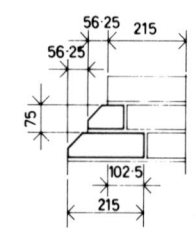

**Detail C**
Scale 1:20

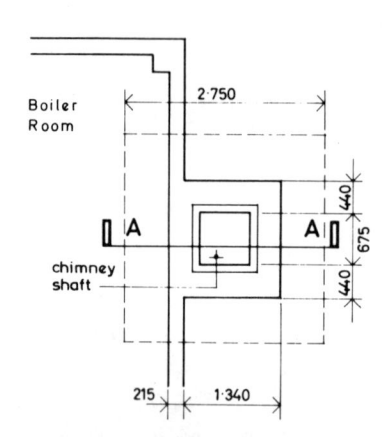

**Plan of Shaft**
Scale 1:100

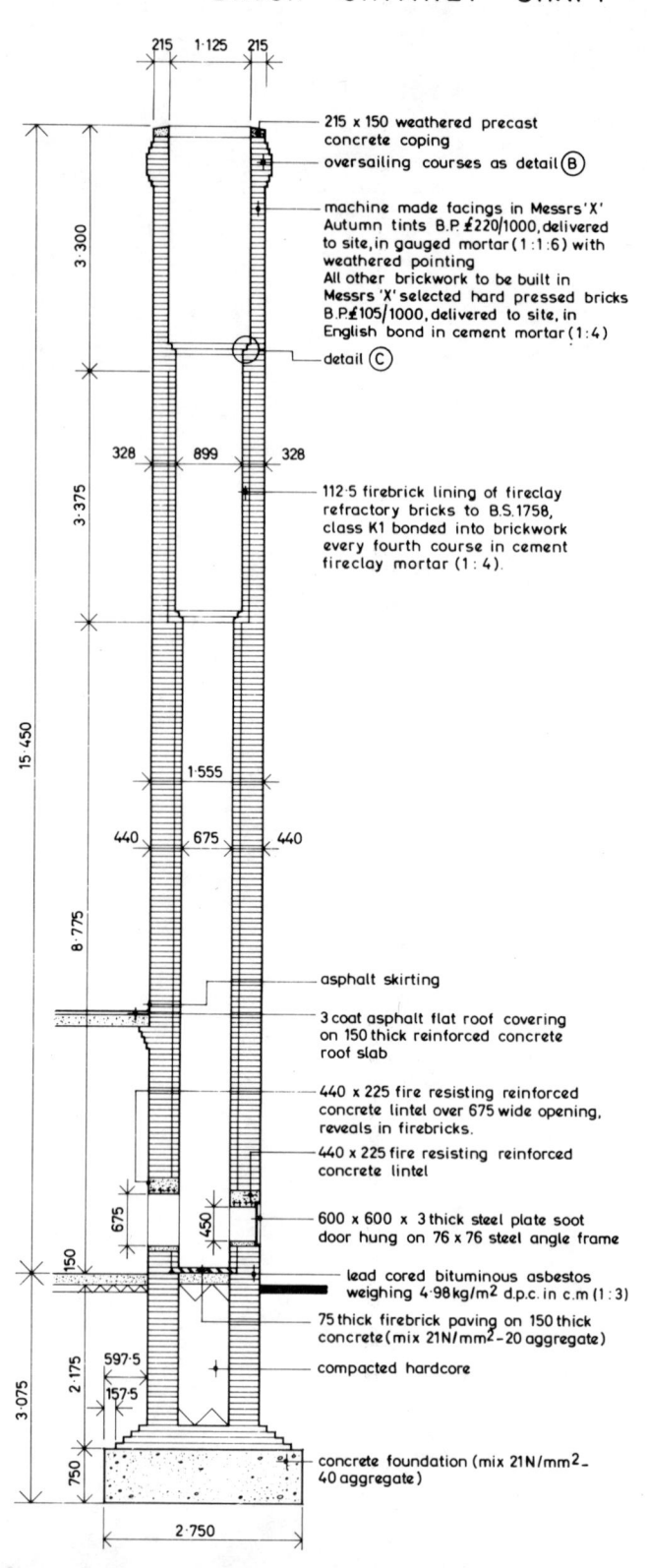

**Section A-A Through Shaft**
Scale 1:100

- precast concrete coping
- 215 x 150 weathered precast concrete coping
- oversailing courses as detail B
- machine made facings in Messrs 'X' Autumn tints B.P. £220/1000, delivered to site, in gauged mortar (1:1:6) with weathered pointing. All other brickwork to be built in Messrs 'X' selected hard pressed bricks B.P. £105/1000, delivered to site, in English bond in cement mortar (1:4)
- detail C
- 112·5 firebrick lining of fireclay refractory bricks to B.S.1758, class K1 bonded into brickwork every fourth course in cement fireclay mortar (1:4).
- asphalt skirting
- 3 coat asphalt flat roof covering on 150 thick reinforced concrete roof slab
- 440 x 225 fire resisting reinforced concrete lintel over 675 wide opening, reveals in firebricks.
- 440 x 225 fire resisting reinforced concrete lintel
- 600 x 600 x 3 thick steel plate soot door hung on 76 x 76 steel angle frame
- lead cored bituminous asbestos weighing 4·98 kg/m² d.p.c. in c.m (1:3)
- 75 thick firebrick paving on 150 thick concrete (mix 21 N/mm² – 20 aggregate)
- compacted hardcore
- concrete foundation (mix 21 N/mm² – 40 aggregate)

Boiler Room

chimney shaft

## BRICK CHIMNEY SHAFT

The follg. in 1nr. sq. bk.chy.shaft, 1.555 m × 1.555 m on plan & 15.60 m o/a ht. above e.g.l. to be built from outside scaffolding.

<u>Substructure</u>
<u>Excavn.</u>

|  |  |
|---|---|
| 2.75<br>2.75 | Exc. topsoil for preservn. av. 150 deep. |
| 2.75<br>2.75<br>0.15 | Disposal of excvtd. mat. on site in spoil hp. av. dist. of 20 m from excavn. |

|  |  |
|---|---|
| bwk. | 2.175 |
| fdn. | 750 |
|  | 2.925 |
| less topsoil | 150 |
|  | 2.775 |

|  |  |
|---|---|
| 2.75<br>2.75<br>2.78 | Exc. pit, max. depth ≤ 4.00 m.   (1 nr)<br>&<br>Fillg. to excavns. av. thickness > 0.25 m, arisg. from excavns. |
| 2.75<br>2.75 | Compactg. bott. of excavn. |
| 4/ 2.75<br>2.93 | Earthwk. suppt., max. depth ≤ 4.00 m, dist. between opposg. faces 2.00 – 4.00 m. |

This has been taken as an isolated chimney shaft and given under an appropriate heading, stating the number, size on plan, shape and overall height, and whether built from outside scaffolding (SMM F10.7.1.1.1).

Excavation work is measured to give the student increased practice in the measurement of this class of work as it raises several additional aspects.
A superficial topsoil excavation item as SMM D20.2.1.1.0 followed by a cubic removal item as SMM D20.8.3.2.1.
When the measurement of the one brick enclosing wall to the boiler room is taken, the necessary adjustments will need to be made for the chimney shaft. Alternatively the chimney shaft dimensions could be adjusted for the one brick wall.
Calculate excavation depth for chimney shaft base measured below the topsoil, which has already been taken.

The pit classification is found in SMM D20.2.4.4.0, and no working space is needed as the requirement in SMM D20.M7 does not apply.

The excavated soil is taken as filling in the first instance to be adjusted as remove from site, when measuring the concrete and brickwork.
Surface treatment of bottoms of excavation is measured in m² (SMM D20.13.2.3.0).
Earthwork support is measured to all four faces of pit in m² without any reference to location except in special cases, but stating the maximum depth range and the distance between opposing faces (SMM D20.7.3.2.0).

4.1

## EXAMPLE 4

|  |  |
|---|---|
| add fdn. proj. 2/597.5 | 2.750<br>1.195<br>3.945 |

|  |  |
|---|---|
| 3.95<br>0.15 | Ddt. ditto.   (topsoil inside boiler rm. |
| 2.75<br>2.75<br>0.75 | In situ conc. isoltd. fdns. 21 N/mm² – 40 agg. poured on or against earth.<br>&<br>Ddt. Fillg. to excavns. a.b.<br>&<br>Add Disposal of excavtd. mat. off site. |

<u>Bk. ftgs.</u>

|  |  |
|---|---|
|  | 2.750 |
| less conc. projs. 2/157.5 | 315 |
| len. of bott. cos. of ftgs. | 2.435 |
| less base of chy. shaft | 1.555 |
| total proj. | 2) 880 |
| proj. ea. side | 4) 440 |
| proj. ea. cos. | 110 |

|  |  |
|---|---|
| mean len. of ea. proj. cos. |  |
| base of shaft | 1.555 |
| add 2/½/110 | 110 |
| top cos. | 1.665 |
| add 2/110 | 220 |
| 2nd cos. | 1.885 |
| add 2/110 | 220 |
| 3rd cos | 2.105 |
| add 2/110 | 220 |
| bott. cos | 2.325 |

Bk. wallg. in Messrs. 'X' sd. hd. pressd. common bks. B.P. £105/1000, delvd. to site in Eng. bond in c.m. (1:4) in fdns.

The earthwork support for the depth of the topsoil excavated inside the boiler room is deducted, although it is recognised that further adjustments will be necessary when measuring the work associated with the one brick wall.
The mix or strength of the concrete is stated in accordance with SMM E10.51. The classification of isolated foundations is used as they will be detached from the wall foundations. SMM7 aims to further rationalise descriptions and so no mention is made of the chimney shaft.
Soil disposal adjustment follows.
It is assumed that soil and groundwater conditions are described on the location drawings for the main building and the general excavation work will include an item for disposal of surface water.
The calculation of the mean length of the projection of each course taken beyond the face of the chimney shaft is calculated in waste. In the first instance it is necessary to determine the projection of each course on each side of the shaft, which is 110 mm and is approximately equivalent to half-a-brick.
To avoid repetition of the description of the bricks, bond and mortar in the description of each item (SMM F10.S1-3), these are all incorporated in a heading, adopting the same procedure that will almost certainly be adopted in the Bill.
B.P. represents basic price of bricks.
It is advisable to state that the work is in foundations, as recommended by the Code of Procedure.
The size of bricks (usually 215 × 102.5 × 65 mm) is normally given in a preamble clause.

4.2

## BRICK CHIMNEY SHAFT (Contd.)

| | | | |
|---|---|---|---|
| 4/ | 1·67 | Bk. projs., width: 110, depth: 75, hor. | |
| 4/ | 1·89 | Ditto., width: 220, depth: 75, hor. | |
| 4/ | 2·11 | Ditto., width: 330, depth: 75, hor. | |
| 4/ | 2·33 | Ditto., width: 440, depth: 75, hor. | |

To arrive at the total mean girth of the projection beyond the shaft face on each course, the mean length for each side is timesed by four.
The brickwork in projections is measured in metres, stating the width and depth of projection (measured beyond the face of the shaft) and the plane (SMM F10. 5.1.3.0).
Each projecting course of footings is given separately, so that the contractor can assess the cost of any rough cutting involved.

| 1·56 |
| 0·30 |

Bk. wall, thickness: 1·555 m, hd. pressed bks.

The brick base to the shaft is measured in m² stating the thickness (4 courses high) as SMM F10.1.1.1.0. Work is deemed vertical unless otherwise described (SMM F10. D3).
By calculating the mean thickness of the brick base it is possible to restrict the measurements to a single item.
It is logical and convenient to proceed with the adjustment of soil disposal for the volume occupied by the brick base and footings to the chimney shaft.

```
        Soil adjustment
                        1·555
   add mean projs.
   2/114 B = 2/280       560
   mean thickness       2·115
```

| 2·12 |
| 2·12 |
| 0·30 |

Ddt. Fillg. to excavns. a.b.
&
Add Disposal of excavtd. mat. off site.

```
    Bk. wallg. below dpc
                        depth
                        2·175
        less ftgs.       300
                        1·875
        add e.g.l. to dpc 150
                        2·025
                        girth
            4/675       2·700
        add corners 4/440  1·760
                        4·460
```

Calculation of the depth of brickwork from the top of the footings up to damp-proof course, from the figured dimensions on the drawing.
Calculation of mean girth of 2B wall from internal dimensions and additions for the four corners.

| 4·46 |
| 2·03 |

Bk. wall, thickness: 440, hd. pressed bks.

Alternatively, the brick wall could be described as 2B thick.

4·3

---

Bk. wallg. w. facewk. o.s. in machine made fcgs. in Messrs. 'X' autumn tints B.P. £220/1000, delvd. to site, in Eng. bond, in g.m. (1:1:6) w. wethd. jts.

```
   e.g.l. to dpc    150
   one cos. below e.g.l.  75
                   225
```

The heading shall include the kind and quality of facing bricks, type of bond, composition and mix of mortar, and type of pointing (SMM F10. S1-4).

The description of the bricks and mortar has been given already in a heading and does not need repeating.
Brick walling with facework on one or both sides is separately measured and described.
Faced brick walling adjusted for that in commons measured from damp-proof course to one course below ground level to allow for any irregularities in the ground surface.

| 4·46 |
| 0·23 |

Add Bk. wall, thickness: 440, facewk. o.s.
&
Ddt Bk. wall, thickness: 440, hd. pressed bks.

### Hdcore.

| 0·68 |
| 0·68 |
| 1·88 |

Fillg. to make up levs., av. thickness > 0·25 m, obtnd. off site, sel. gravel rejects blinded w. hoggin, & compactg. in layers, thickness ≤ 250.

Filling to make up levels, classifying the thickness and describing the source, type of filling and specified handling details as SMM D20.10.2.3.4.

| 0·68 |
| 0·68 |

Compactg. fillg. blinded w. hoggin.

Treating the surface of filling is described and measured in m² as SMM D20.13.2.2.1.

```
                    1·875
        less topsoil  150
                    1·725
```

Calculation of depth of brickwork and hardcore for soil disposal.

| 1·56 |
| 1·56 |
| 1·73 |

Ddt. Fillg. to excavns. a.b.
&
Add Disposal of excvtd. mat. off site.

The adjustment of soil disposal follows the measurement of brickwork and hardcore filling.

```
        2/1·340     2·680
                    1·555
   add 2 corners 2/597·5  1·195
                    5·430
```

| 5·43 |
| 0·60 |
| 0·15 |

Fillg. to excavns., av. thickness ≤ 0·25 m, obtnd. from on site spoil heaps, topsoil.

The only adjustment item then remaining is to return the topsoil around the outside of the shaft.
It is measured in m³ in accordance with SMM D20.9.1. 2.3.

4·4

## BRICK CHIMNEY SHAFT (Contd.)

### Base to Shaft

| | | |
|---|---|---|
| 0·68<br>0·68<br>0·15 | In situ conc. bed, 21N/mm² – 40 agg., thickness ≤ 150. | Concrete beds are measured in m³, giving the thickness classification as SMM E10.4. 1.0.0. As it is laid on blinded hardcore, this does not have to be mentioned. |
| 0·68<br>0·68 | Trowellg. surf. of conc. | Trowelling the surface of concrete is a separate superficial item in accordance with SMM E41.3.0.0.0. |

```
          add  2/102·5    675
                         205
                         880
```

| | | |
|---|---|---|
| 0·88<br>0·88 | Firebk. pavg., 300 × 300 × 75th. w. smth. fin. b. & j. in fire ct. mo. (1:4), laid level on conc. base to shaft (m/s). | The firebrick paving to the base of the shaft must be adequately described, including the plane, kind and quality of materials, size, thickness, nature of base and method of bedding (SMM M40.5.1.1.0 and M40.51–7). |

### Dpc

| | | |
|---|---|---|
| 4·46<br>0·44 | Dpc, width > 225, hor., of single layer of lead cored bit. asbestos weighg. 4·98 kg/m² to BS 747, & bedded in c.m. (1:3). | The description of the damp-proof course is to include the kind, quality and substance of damp-proof material, number of layers, and composition and mix of bedding material (SMM F30. 51, 4–6), measured in m² stating the appropriate width range and the plane as SMM F30. 2.2.3.0.<br>Pointing exposed edges are deemed to be included without the need for specific mention (SMM F30.C2). |

### Superstructure

```
          4/1·555      6·220
       less corners 4/328   1·312
                            4·908
```

| | | |
|---|---|---|
| 4·91<br>8·78 | Bk. wall, thickness: 328, facewk. o.s., a.b., bldg. against other wk. (bott. stage) | Proceed to measure the brickwork up the chimney shaft, following the order of construction on site. It is not necessary to describe the bricks, bond and mortar, as these have been described previously and so the letters 'a.b.' have been inserted.<br>Note that the wall is 1½ bricks thick and not 2 bricks as the dimension of 440 includes the firebrick lining and this brick walling item includes the facework and building against the firebrick lining (SMM F10.1.2. 1.1.) |

4·5

```
                        ht.
                       8·775
    less firebk. paving   75
                       8·700
                        gth.
         4/675        2·700
    add corners 4/112·5  450
                       3·150
```

| | | |
|---|---|---|
| 3·15<br>8·70 | Flue ling., thickness: 112·5, fireclay refractory bks. to BS 1758, class K1, bonded into gen. bwk. every 4th. cos. in stret. bond & backed & jtd. in fire c.m. (1:4); fin. fair w. flush. jts. as wk. proceeds. (bott. stage) | The dimensions of the lower length of firebrick lining are calculated from the figured dimensions and entered in waste. The half-brick firebrick lining is taken as 112·5 mm thick to allow for the 10 mm thick mortar joint between the firebricks and the hard pressed bricks.<br>The firebrick lining is measured and described in accordance with SMM F10.51–4 and given separately in m² stating the thickness (SMM F10.9.1.0.0). There will be a 10 mm thick mortar joint at the back of the lining. |

```
                             6·220
         less corners 4/215   860
                             5·360
```

| | | |
|---|---|---|
| 5·36<br>3·38 | Bk. wall, thickness: 215, facewk. o.s., a.b., bldg. against other wk. (2nd stage) | The second stage of the shaft consists of a 1B wall enclosing a ½ B firebrick lining. The girth of the one-brick wall is calculated from the outside girth by making deductions for the corners.<br>As a check, a calculation can be made working from the outside of the shaft when a slight variation in girth occurs because of the half millimetres involved. A 328 wall is really 327·50 mm thick. |

```
                            ht.
                          3·375
   less bott. 2 p.m. spld.
        header courses     150
                          3·225
                           gth.
          4/899           3·596
   add corners 4/112·5    450
                          4·046
```

```
   Thus:  ex. girth        6·220
   less 4/2/215   1·720
        4/112·5    450    2·170
                          4·050
```

| | | |
|---|---|---|
| 4·05<br>3·23 | Flue ling., thickness: 112·5 a.b. (2nd stage) | The full description need not be repeated as it has already been given in full in a previous entry. Calculation in waste of mean girth of splayed header course of firebricks. |

```
          4/899 = 3·596
```

| | | |
|---|---|---|
| 3·60 | Flue ling., plinth cappg. of pm spld. firebks., projectg., depth of set fwd: 56, width: 215, hor., entirely of headers. | This item is measured in metres stating the width and depth of projection, and that it consists entirely of headers as for splayed plinth cappings in SMM F10.13. 3.3.3 and F10.D11. Labours in angles are deemed to be included (SMM F10.C1f). |

4·6

## BRICK CHIMNEY SHAFT (Contd.)

| | | | |
|---|---|---|---|
| | 3·596 | | |
| | add corners 4/56·25  225 | Calculation of girth of upper splayed header course of firebricks. | |
| | 3·821 | | |
| | 215 | | |
| | 56 | | |
| | 159 | | |
| 3·82 | Flue ling., plinth cappg. p.m spld. firebks., projectg., depth of set fwd: 56, width: 159, hor., entirely of headers. | Linear upper splayed header course item of smaller overall width but same projection. | |
| | 3/75 = 225 | Calculation of height of 3 courses of brickwork to 1½ brick wall above firebrick lining. The girth has been calculated previously. | |
| 4·91 | Bk. wall, thickness: 328, facewk. o.s., a.b. (above firebk. ling.) | | |
| 0·23 | | | |
| | 4/1·125 = 4·500 | Calculation of mean girth of lower course of splayed header hard pressed bricks. Same procedure as for header course of splayed firebricks. | |
| 4·50 | Bk. plinth capping., pm spld. hd. pressed bks., projectg., depth of set fwd: 56, width: 75, total depth: 215, hor., entirely of headers. (base of top stage of shaft) | | |
| | 4·500 | | |
| | add corners 4/56·25  225 | | |
| | 4·725 | | |
| 4·73 | Bk. plinth capping., pm spld. hd. pressed bks., projectg., depth of set fwd: 56, width: 75, total depth: 102·5, hor., entirely of headers. | Upper course of splayed bricks at base of upper section of one-brick walls. | |
| | bott. cos. | | |
| | ext. gth. 6·220 | | |
| | less corners 4/102·5  410 | This leaves two courses to measure on the outer face of the shaft adjoining the special header courses. One is half-brick thick and the other three quarter brick thick. | |
| | 5·810 | | |
| | top. cos. | | |
| | ext. gth. 6·220 | | |
| | less corners 4/155  620 | | |
| | 5·600 | | |
| 5·81 | Bk. wall, thickness: 102·5, facewk. o.s., a.b. (adjg. spld. hor. cos.) | Both courses are measured in m² stating the thickness (SMM F10.1.2.1.0). | |
| 0·08 | | | |
| 5·60 | Bk. wall, thickness: 155, facewk. o.s., a.b. (do.) | | |
| 0·08 | | | |

4.7

| | | | |
|---|---|---|---|
| | ht. 3·300 | Calculation of height of top section of one-brick wall. The mean girth has been calculated previously. | |
| | less copg.  150 | | |
| | 2 hdr. cos. 150 | | |
| | 1½B wall  225  525 | | |
| | 2·775 | | |
| 5·36 | Bk. wall, thickness: 215, facewk. o.s., a.b. (top sec. of shaft.) | Adjustment for facework inside the boiler room and adjoining one-brick walls to be taken with the boiler room. | |
| 2·78 | | | |
| | Projectns. | Projections in oversailing courses are measured in metres stating the width and depth of projection and plane (SMM F10. 5.1.3.0). | |
| | ext. gth. 6·220 | | |
| | add corners 4/28  112 | | |
| | ¢ of bott. cos. 6·332 | | |
| | add corners 4/28  112 | | |
| | ¢ of top cos. 6·444 | | |
| | add corners 4/28  112 | | |
| | ¢ of projg. bd. 6·556 | The top course of the three is taken as part of the projecting band above. The girths are measured on the centre line of each projection and the depth taken from the outer face of the shaft wall. | |
| 6·33 | Bk. proj., facewk., width: 75, depth of proj. 28, hor. | | |
| 6·44 | Ditto., width: 75, depth of proj. 56, hor. | Labours in returns and angles are deemed to be included (SMM F10.C1f). | |
| 6·56 | Ditto., width: 225, depth of proj. 84, hor. | | |
| | Facewk. ornamentl. bands | Calculation of girths of splayed plinth cappings measured on external faces. | |
| | 6·220 | | |
| | add corners 4/2/28  224 | | |
| | 6·444 | | |
| | top course add 4/2/28  224 | | |
| | 6·668 | | |
| | middle cos. add 4/2/28  224 | | |
| | bott. cos. 6·892 | | |
| 6·44 | Bk. plinth cappgs., pm spld. fcg. bks. projectg., depth of set fwd: 28, width: 75, hor., e.o. wk. in which they occur, entirely of headers. | The facework to the splayed plinth capping courses to the ornamental band is measured in metres giving the relevant particulars listed in SMM F10. 13.3.3.1 & 3) and F10.D11. | |
| 6·67 | | | |
| 6·89 | | | |
| | Ex. bwk. at plinth cappgs. | | |
| 6·67 | Bk. wall, thickness: 56 in hd. pressed bks. | Average thickness taken over three plinth capping courses from outer face of shaft. | |
| 0·23 | | | |

4.8

## BRICK CHIMNEY SHAFT (Contd.)

### Corbel to support roof slab

| | | |
|---|---|---|
| 1·56 | Bk. projs., facewk., width: 300, ov. depth of proj: 75, hor. | This item picks up the additional faced brickwork in the four oversailing courses on one face of the shaft to support the roof slab to the boiler room (SMM F10.5.1.3.0). It is not considered necessary to adjust the faced brickwork to the chimney shaft, as the area is so small. |

### Coping

| | | |
|---|---|---|
| 5·36 | Precast conc. copg., spld., 215 × 150 o/a to BS 3798, incorptg. ord. P. ct., b. & j. in c.m (1:3). (In 4 nr units) | The mean girth of the coping has been calculated previously for the one-brick wall. This is a linear item as SMM F31.1.2.0.0, giving the number of units and the particulars listed in SMM F31.S1 & 4. |
| 4/  1 | E.O. copg. for ∟s. | Angles taken as extra over enumerated items as SMM F31.2.1.0.0. |

Finally adjust for the two openings at the base of the shaft.

```
                Openings
                 width
  add. firebk.   450    675
  lings. 2/112·5 225    225
                 675    900
                        ht.
                 450    675
  add lintel 225
  f.b. sill   75 300    300
                 750    975
```

The areas for the deduction of faced brickwork and firebrick lining are first calculated in waste.

These openings are too large for the provisions of SMM F10.M2a to apply.

| | | |
|---|---|---|
| 0·68 0·75 | Ddt. Bk. wall, thickness: 328, facewk. o.s., a.b. & | Deduction of the two superficial items with abbreviated descriptions and suitably bracketed. |
| 0·90 0·98 | Ddt. Flue ling. thickness: 112·5, a.b. | |
| 0·68 0·90 | Firebk. sill, 440 wide × 75 dp., hor. b. & j. in fireclay c.m. (1:4) & fin. w. flush jts. | The firebrick sills are measured as linear items and giving the particulars listed in SMM F10.15.1.3.0. Labours in ends are deemed to be included (SMM F10.C1f). |
| 2/ 0·44 0·45 2/ 0·44 0·68 | Firebk. ling. to opgs., thickness: 112·5, facewk. o.s., in f.b. refractory bks. a.b. & bldg. against other wk.  (reveals) | Measured in accordance with isolated casings (SMM F10.3.2.1.1) as being the nearest SMM7 item, and giving additional particulars to assist in identification as SMM General Rules 1.1. |

4·9

```
                        450    675
          add ends 2/225 450    450
                        900   1·125
```

| | | |
|---|---|---|
| 1 | Precast conc. lintel, 440 × 225 × 1·125 lg. in fire resistg. conc. (1:2:4/20agg.) reinfd. w. 4 nr 16 ⌀ m.s. bars to BS 4449, w. 2 sides & soff. fin. fair & bedded in fireclay c.m. (1:4). | Add bearings to width of openings to give length of lintels. Precast concrete lintels are enumerated with a dimensioned description and giving the particulars listed in SMM F31.1.1.0.1 and F31.S1-5. Moulds are deemed to be included with the item (SMM F31.C1). |
| 1 | Precast conc. lintel 440 × 225 × 900 lg. a.b., but reb. | Similar construction as previous item, so the letters a.b. are used to avoid the repetition of a long description. |
| 1 | Soot dr. 600 × 600 × 3th. m.s. plate, approx wt.- kg., hung on 76 × 76 × 9 m.s. ∟ fr. welded at jts. & fxd. to f.b. ling. & bk. wall 440 wide w. stand stl. lugs, & b. & j. in fireclay c.m. (1:4). Dr. & fr. to be primed w. ② cts. of cal. plumb. primer at supplier's wks., after fabricatn. Strap hinges & fasteng. device as component dwg. M2A. | Enumerated item for soot door, frame and fixing giving the various particulars listed in SMM F30.14.1.0.0. Any painting as part of the production process can be included in the description. It would be permissible to include the forming of the opening, lintel and liners in the description of the enumerated soot door item (SMM F30.C8), but is seems preferable to deal with both openings in a similar manner in this instance. |
| 1 | Paintg. gen. isoltd. surfs, met., area ≤ 0·50m², prime & ③. & Ddt., ext.  (soot dr. | This item has been enumerated in accordance with SMM M60.1.0.3.0 as the area of the door is less than 0·50m², and giving the particulars listed in SMM M60.S1-6. |

To take

Adjustments to faced brickwork to shaft in boiler room.

Notes of any outstanding items should also be inserted.

4·10

## DRG. 5 — MASONRY

**NOTES:**
Stone to be Whit bed Portland stone, rubbed on exposed faces and laid with natural bed at right angles to load.
Mortar to be 1 part Portland cement, 3 parts lime, 12 parts crushed stone dust.

- cavity tray
- face of voussoirs

### HALF ELEVATION
Scale 1:50

### VERTICAL SECTION
Scale 1:50

### B — SECTION — STONE BALUSTRADE
(Bath Stone) 7·00m length
Scale 1:20

- 280 x 100 coping in 560 lengths
- 25 ⌀ x 50 phosphor bronze dowels
- 150 x 150 x 560 balusters at 280 centres
- 225 x 80 plinth capping dowels under each third baluster
- 440 x 200 cornice in 840 lengths
- two dowels to each length of cornice

### STRUCTURAL DETAILS
Scale 1:20

- reinforced concrete structural wall
- 150 dovetail set in concrete
- 50 x 100 concrete nib
- 13 backing in gauged mortar (1:1:6)
- plain ashlar facing in Darley Dale stone 100 thick
- joggle joint
- fixing cramps in stainless steel L corbel 150 x 75 x 6 overall
- compression bed 13 thick in mastic

### SECTION A-A
Scale 1:20

### SECTION B-B
Scale 1:20

### A — STONE SURROUND TO DOOR OPENING

dpc

### ELEVATION
Scale 1:50

- joints 5 thick
- cement mortar (1:3) joggle joint
- fixing cramps (stainless steel L corbel 150 x 75 x 6 overall) with slot nut preset in dovetail slot in concrete
- 50 x 100 concrete nib
- compression bed
- floor level

### C — PLAIN ASHLAR FACING TO REINFORCED CONCRETE STRUCTURAL WALL

# STONE SURROUND TO DOOR OPENING

| | | | |
|---|---|---|---|
| | St. dressgs of P. st. (Whit bed) rubbed on exp. faces & laid w. nat. bed at right angles to load; bedded; jtd. & ptd. in st. dust mo. (1pt. P.ct., 3pts. lime & 12 pts. crushed st. dust), fin. w. nt. flush jts. all set to proj. 75 from face of bwk. & coatg. back of st. w. lime putty and cleang. down on completn. in st. dr. dressg. to bk. faced wall, as detailed on drg. 5. | A general description of the work, indicating the type of work, stone and mortar to be used as SMM F21.S1–10. Natural stone ashlar walling and dressing consist of dressed blocks accurately worked to given dimensions and laid in mortar with fine joints. Stone dressings are those in walls of other materials (SMM F21.D2). Particulars are to be given of the kind of stone (Portland stone), quality (Whit bed), texture and finish to exposed faces, as Code of Procedure (rubbed), composition and mix of mortar for bedding, jointing and pointing (stone dust mortar – 1:3:12), type of pointing (neat flush joints), and any requirements as to coatings to surface of finished work (lime putty) and cleaning on completion. | |

```
                    steps
  2/910             1·820
      75
     250
     125
  2/450              900
                   2·720
```

| | | |
|---|---|---|
| 2·72 | Step, 300 x 150 plain, > 1·50 m lg.    (1)  (In lnr) | The stones have been numbered on the drawing, prior to taking off, for ease of reference. Length of bottom step is calculated from figured dimensions on drawing, remembering to double all the lengths to cover each side of the door opening. Steps are given in metres with a dimensioned description, stating the number and classifying as plain as opposed to spandrel (SMM F21.21.0.1.0). Blocks > 1·50m long are so described. Labours in returns, ends and angles of stonework are deemed to be included (SMM F21.C1g). The numbers in brackets are the stone identification numbers from the drawing, thus locating each stone in turn and reducing the risk of omission. |

5.1

---

## EXAMPLE 5A

| | | |
|---|---|---|
| 1·82 | Step, 152·5 x 150, plain, > 1·50 m lg.    (2)  (In lnr) | The finish to the exposed faces of the steps is included in the heading and does not need repeating for each item. |

```
         width    Jambs
                thickness
          50      140
         250       75
          75      215
         375
```

| | | |
|---|---|---|
| 2/ 0·30 | Jamb, st., isoltd., 375 wide & 215 th., sk.   (3) | Build up of width and thickness from figured dimensions wherever practicable, or scaling. Jamb stones are measured in metres (vertically) stating whether attached or isolated and giving a dimensioned description including the sinkings (rebates and chamfers) as SMM F21.11.3.1.11. |

```
         width
          125
          250
           75
          450
```

| | | |
|---|---|---|
| 5/2/ 0·23 | Jamb, st., isoltd., 450 wide & 215 th., 2ce sk. & mo.   (4,6,8 10 & 12) | Mouldings are included in the description (SMM F21.M7). Isolated stones are those attached to another form of construction, in this case brickwork (SMM F21.D13). Five jamb stones on each side of the opening, as illustrated on section A–A are identical and can all be taken together in a single item, giving the stone numbers in waste for identification purposes. The rebates and chamfers are all grouped together as sunk in the description, stating the number of times it is sunk (Code of Procedure F21.1-32 – 11). |
| 2/2/ 0·30 | Jambs, st., isoltd., 375 wide & 215 th., 2ce   (5 & 9) sk. & mo. | These two pairs of jamb stones on each side of the door opening are of identical cross section, although varying in length (vertically) and can thus be included in the same billed item. |
| 2/2/ 0·15 | | (7 & 11) |

### Special purpose stones

See text for masonry definitions.

| | | |
|---|---|---|
| 2/ 1 | Springer, 395 x 215 x 300 to semi-circ. arch., mo., 2ce sk. & bldg. against other wk.   (13) | Springers and voussoirs (arch stones) are enumerated stating the function and giving a dimensioned description as SMM F21.33.1.2.0. |

5.2

STONE SURROUND TO DOOR OPENING (Contd.)

| | | | |
|---|---|---|---|
| 2/ | 1 | Voussoir, 400 x 215 x 230 to semi-circ. arch, mo., 3 times sk. & bldg. against other wk. (14) | The natural bed will be on the line of the lowest joint (giving one sunk joint only) and the sizes represent the smallest block from which they can be obtained (SMM F21.M11) as shown in the accompanying sketch. |
| | | | Stones are deemed to be set on their natural beds unless otherwise described (SMM F21.59). The order of length, thickness and width on face has been adopted for each stone. All stones except the springer have a radiating bed and are twice sunk jointed (on both edges). The enclosing rectangle for each voussoir is plotted on the drawing from which the length and width dimensions are scaled. |
| 2/ | 1 | Voussoir, 440 x 215 x 180 to semi-circ. arch, mo., 3 times sk. & bldg. against other wk. (15) | |
| 2/ | 1 | Voussoir, 550 x 215 x 290 to semi-circ. arch, mo., 4 times sk. & bldg. against other wk. (16) | Apart from the keystone there are two identical arch stones on each side of the arch and so each stone is twiced in the timesing column. There are no clear directions in SMM7 as to what the dimensioned descriptions for special purpose stones should include. Hence the procedure outlined for other natural stonework has largely been followed, although it must be emphasised that this is only one of several possible approaches. The term 'sunk' refers to a plain face worked below the surface of the stone, such as chamfers and rebates, and it is necessary to state the number of times it is sunk. The term could not therefore embrace mouldings which need to be specifically mentioned in the description. |
| 2/ | 1 | Voussoir, 700 x 215 x 400 to semi-circ. arch, mo., 4 times sk. & bldg. against other wk. (17) | |

5.3

| | | | |
|---|---|---|---|
| 2/ | 1 | Voussoir, 750 x 215 x 360 to semi-circ. arch, mo., 4 times sk. & bldg. against other wk. (18) | Building against other work refers to building against brickwork as a dressing. Another alternative is to refer to dimensioned diagrams for each stone showing the shape and dimensions of the work as SMM General Rules 5.3 and as illustrated in figure 3 of the Code of Procedure. |
| 2/ | 1 | Voussoir, 890 x 215 x 360 to semi-circ. arch, mo., 4 times sk. & bldg. against other wk. (19) | A fully dimensioned description for this item follows to show just how lengthy it would become if all labours and their dimensions were given: 'Voussoir, 890 x 215 x 300 to semi-circ. arch, sk. jtd. 790 wide, stepd. sk. jtd. 680 wide, circ. sk. 130 gth., circ. reb. av. 585 gth., circ. mo. 120 gth., w. rad. bed & bldg. against other wk.' |
| | 1 | Keystone, 1000 x 215 x 640 to semi-circ. arch, mo., 3 times sk. & bldg. against other wk. (20) | The width of the keystone as scaled from the drawing has to be doubled as only one-half of the stone is shown on the elevation. Unlike the other arch stones the keystone is not recessed or rebated, or stepped sunk jointed. |
| | 1 | Centrg. to arch, semi-circ. st., 1.82 m span, 215 wide on soff. & 910 rise, max. suppt. ht: 3.00-4.50 m. | Enumerated item for centering as SMM F21.36.1.1.2, giving a dimensioned description. Where the maximum support height is 3.00 m or more, it is necessary to state the height in 1.50 m stages. The items to be included in the dimensioned description are listed in SMM F21.M12. |

5.4

| | | STONE BALUSTRADE | | | | EXAMPLE 5B | |
|---|---|---|---|---|---|---|---|
| | | (7.00 m length) Bath st. (Coombe Down) as sample supplied, rubbed on exp. faces & laid w. nat. bed at rt. L's to load; bedded, jtd. & ptd. in ct: lime: sd. (1:3:12), fin. w. nt. flush jts. & cleang. down on completn. in st. balustrade, & fixg. ea. baluster to the copg. & ea. 3rd. baluster to the plinth cappg. w. phosphor bronze dowels, 25ϕ x 50 lg. | It is advisable to start with a general description of the work including particulars of the stone, finish, mortar, pointing and fixing as SMM F21.S1-10. | | 7·30 | Band, cornice, 400 x 200, hor., weathd. & mo., to profile as component drg. 5B. | Same measurement rules as for the capping and coping, but with the intricate moulded face it is necessary to refer to the component drawing, or alternatively to supply or insert a dimensioned diagram. Cornices come under the heading of bands (SMM F21.14.1.3.0 and F21.D15). |
| | | add proj. ends 2/150    7·000<br>                     300<br>                     7·300 | | | | | |
| | 7·30 | Copg., 280 x 100, hor., wethd., circ. reb. & 2ce thro., w. joggle jts. | Copings are measured in metres with a dimensioned description and stating the plane as SMM F21.16.1.3.0. Labours in ends are deemed to be included (SMM F21.C1g). | | | | |
| | | 280 ) 7·000<br>       25 + 1 | | | | | |
| 26/ | 1 | Balusters, ex 150 x 150 x 560, fully circ. mo. as component drg. 5B. | Balusters would be classified as special purpose stones and enumerated with a dimensioned description as SMM F21.33.1.2.0.<br>Because of the complexity of the profile it is desirable to refer to a component or dimensioned diagram. Metal dowels and their associated mortices are deemed to be included in the stonework (SMM F21.C1c & e). | | | | |
| | 7·30 | Cappg., 225 x 80, hor., mo. (plinth | Cappings are measured in metres, giving a dimensioned description and the plane as SMM F21.18.1.3.0.<br>Mouldings are given in the descriptions of linear items (SMM F21.M7). | | | | |
| | | | 5·5 | | | | 5·6 |

EXAMPLE 50

## PLAIN ASHLAR FACING

Plain ashlar to front face of 3 storey office building 12 m. hi. of reinforced concrete walls, as location drawing NR/23/2.

| | | |
|---|---|---|
| | | It is helpful to provide a general description of the work preceding its measurement. |
| | | A small area of 2·50 m x 2·10 m is taken for the purpose of this example, to illustrate the approach to the measurement of the facade as a whole. |
| 2·50<br>2·10 | Ashlar, plain st. dressg. to wall, Darley Dale st., thickness: 100, from 'X' quarry, in slabs 500 x 700 w. lg. side vert. & faced o.s. w. tooled fin., built against other wk. w. g.m. (1:1:6) backg., 13 th. & fxg. w. 2nr fixg. cramps/slab of stainless stl. L corbel 150 x 76 x 6 o/a holed, to fit over slot nut preset in dovetail slot in conc. (m/s), w. c.m. (1:3) joggle jts. to vert. edges & bedded in g.m. (1:1:6) w. 5th. jts., fin. smth. & cleang. on completn. | The ashlar slabs are measured in m² in accordance with SMM F21.1.1.1.3, 4, 6 & 10, and giving the particulars listed in SMM F21.S1-12. The description is to include details of the method of fixing of the slabs, in addition to the type of stone and its finish, coatings to backs of stones, composition and mix of mortar, type of pointing and method of jointing. All rough and fair square cutting, metal cramps, dowels and the like are deemed to be included (SMM F21.C1e & l). Quoin stones are classified as attached or isolated and measured in metres with a dimensioned description (SMM F21.10.1-3.1.0). |
| 2·50 | Reb. 50 x 100. (over nib in r.c. wall<br><br>&<br><br>Compressn. jt., in st. dressg. 100 wide & 13th. in mastic to BS 3712. | An additional linear item is taken to pick up the rebate, stating the size (SMM F21. 31.1.0.0).<br><br>A further linear item is included to cover the compression joint, which would otherwise go unmeasured and could then give rise to a claim by the Contractor for an extra. |

5.7

# 4 REINFORCED CONCRETE

## CATEGORIES

The student will be aware of the significance of reinforced concrete in which steel reinforcement in the form of bars or fabric is inserted to provide much greater resistance to tensile stresses. It is necessary to distinguish between *in situ* concrete, where the concrete is poured in its final position on the site, and precast concrete, where the components are prefabricated and subsequently placed in position in the structure. Yet another type is prestressed concrete, where a stress is artificially induced in the concrete by means of tendons tensioned before working loads are applied. Gun applied concrete is measured in m$^2$, with slabs, walls, beams and columns separately classified (SMM E11.1–4.1.0.0).

## MEASUREMENT OF REINFORCED CONCRETE

The concrete work in the Bill should start with a general description of the work, where it is not evident from the location drawings. Although the location drawings or further drawings accompanying the bill of quantities should show the relative positions of concrete members, size of members, thickness of slabs and permissible loads in relation to casting times (SMM E10.P1). The concrete is classified according to the types of members of which it forms part.

The normal procedure in measuring a reinforced concrete structure is to work on a floor by floor basis, proceeding from the lowest floor upwards, and making the necessary adjustments on walls, columns, and floor and roof slabs for formwork to beams. Reinforced concrete work consists of three component parts — concrete, reinforcement and formwork — and it is important to ensure that all three are measured for each part of the structure.

The bulk of *in situ* concrete work is measured by volume, stating the thickness range for beds, slabs and walls.

Isolated columns are classified as columns or walls, according to the ratio of length to thickness; to be measured as an isolated column, the length on plan is ≤ four times the thickness (SMM E10.M4).

*In situ concrete*

All concrete must be adequately decribed by kind and quality of materials and mix or strength (SMM E10.S1). Reinforced work shall be so described and, where the volume of reinforcement > 5 per cent of that of the concrete, this shall be stated because of the difficulties of placing and compacting concrete in heavily reinforced members. The Code of Procedure emphasises that this high reinforcement content will only apply in exceptional cases, as it will have to exceed 0.41 t/m$^3$ of the measured member.

Unless concrete in foundations, ground beams and the like is poured on or against earth or unblinded hardcore and so described, formwork must be measured to support the concrete.

An extensive list of concrete categories is incorporated in SMM E10.1–14, where a considerable number of groupings occur. For example, concrete to foundations is deemed to include attached column bases and pile caps (SMM E10.D1). Slabs and their supporting beams are measured in m$^3$, stating the thickness of the slab in the appropriate thickness range as SMM E10.5.1–3. The thickness range is not varied for the attached beams and beam casings which are included in with the volume of the slab, provided their depth is ≤ three times their width (depth measured below the slab). Similar approaches are applied to beds, blinding beds, plinths and their thickenings (SMM E10.D3), and walls with attached columns and piers (SMM E10.D6).

When measuring treating the surface of concrete, each type such as power floating and trowelling is kept separate, distinguishing between finishing sloping, to falls, crossfalls and soffits and measured in m² (SMM E41.1–7.0.0). Leaving the top surfaces of concrete foundations to receive brickwork does not fall into this category.

*Reinforcement*

The two main types of reinforcement are measured differently. Bars are measured by length at taking off stage and subsequently billed in tonnes to two places of decimals (SMM E30.1 and General Rules 3.3). Fabric reinforcement is measured in m², although strips in wall foundations and tension strips to floors and roofs are separately described stating the width of strip (SMM E30.4.1.0.2).

Each diameter (nominal size) of bar is given separately, as also are straight, bent and curved bars (SMM E30.1.1.1–3), and the billed item is deemed to include hooks and tying wire, and spacers and chairs which are at the discretion of the contractor, without the need for specific mention (SMM E30.C1). Reinforcement is not classified as to location. The description of the bars shall state the kind and quality of materials (SMM E30.S1). Since it is relatively difficult to handle and fix long bars, horizontal bars and bars sloping ≤ 30° from the horizontal 12.00 m long and over are so described, stating the length in further stages of 3.00 m, while with vertical bars and those sloping > 30° from the horizontal, the separate 3.00 m stages start with lengths of 6.00 m and over. Bent bars cover those which are specifically bent to curves to suit the shape of the member in which they are to be placed (Code of Procedure E10.1.1.3).

The length, number and shape of each bar type is generally given in a reinforcement or bar bending schedule, from which the details are extracted at the taking off stage. A check should however be made of all entries and, in the event of any disagreement, the point of difference should be recorded on a query list. Where no reinforcement schedule is provided, the student has to calculate the lengths of the bars from the drawings, using figured dimensions as far as practicable. It will be necessary to make adjustments for concrete cover to the reinforcement to prevent rusting — usually taken at about 40 mm— and additions for hooked ends (often calculated at twelve times the diameter of the bar) and turned ends (usually around 75 mm).

Lengths of each of the bars are then systematically entered in the dimension column, taking care to ensure that each type of bar is listed and that the correct timesing figures are inserted. Example 6 shows the approach to the measurement of bar reinforcement in the absence of a reinforcement schedule, and example 7 shows the procedure to be followed when a reinforcement schedule is supplied. The weights of the different diameter bars can be obtained from BS 4449, steelwork tables and various other reference books, and the weight in kg/m of the more common sizes of bar is given in example 7.

The description of fabric reinforcement includes the kind and quality of steel (SMM E30.S1). It is measured in m² as the area covered with no allowance for laps and no deductions for voids ≤ 1.00 m² (SMM E30.M4 and 5). The description of the fabric shall state the mesh reference, weight per m² and minimum laps (SMM E30.4.1.0.0 and E30.S6). The estimator has to allow for the laps when calculating his fabric rate/m². Fabric reinforcement is deemed to include laps, tying wire, all cutting and bending, and spacers and chairs which are at the discretion of the contractor (SMM E30.C2).

Some fabric reinforcement is measured to a concrete suspended floor in *Building Quantities Explained* (example X). A typical description would be 'Steel fabric reinforcement to BS 4483 ref. A193, weighing 3.02 kg/m², with 100 minimum laps'.

*Formwork*

Formwork is grouped in one of the categories listed in SMM E20.1–26. No deductions are made for voids in formwork to soffits of slabs or walls ≤ 5.00 m², irrespective of location (SMM E20.M4 and 8). Certain categories of formwork are kept separate and suitably described — left in, permanent, and to curved surfaces stating the radii (SMM E20.M2). Formwork to soffits of slabs and landings is classified according to the thickness of the concrete and the soffit height as ≤ 1.50 m high and thereafter in 1.50 m stages. Formwork to walls at a height > 3.00 m above floor level is kept separate and so described and that to vertical and battered surfaces is separately classified.

Formwork to sides of foundations and ground beams, edges of beds and slabs and associated items is measured in m² where >1.00 m high. It is measured as linear items where ≤ 1.00 m high, in three categories — ≤ 250 mm, 250–500 mm and 500 mm–1.00 m high, and described as plain vertical or giving a dimensioned description (SMM E20.1–7.1–2.1–4.1–2).

Formed finishes to concrete are measured in m² as extra over basic finish, being the finish formed from the principal formwork (SMM E20.20.1–5.0.0 and E20.D11 and Code of Procedure). Formwork is deemed to include all cutting, splayed edges and the like, and adaptation to accommodate projecting pipes, reinforcing bars and the like (SMM E20.C1 and 2).

Formwork to soffits of slabs and landings ≤ 200 mm thick are given separately, and thereafter in 100 mm stages, stating the number of soffits to landings, and whether horizontal, sloping ≤ 15° or > 15°, height to soffit ≤ 1.50 m and thereafter in 1.50 m stages and whether left in or permanent (SMM E20.8–9.1–2.1–3.1–4). Formwork is also measured to top surfaces sloping > 15° (SMM E20.11 and E20.M7).

Formwork to beams, beam casings, columns and column casings are classified as either attached to slabs, attached to walls or isolated, and measured in m$^2$ when of regular shape, stating the shape. When of irregular shape, it is measured in metres, supported by a dimensioned diagram as illustrated in example 7. In both cases it is necessary to state the number of members in each item. Regular shaped members include rectangular, circular, hexagonal or other definable regular shape (SMM E20.D10). The same height to soffit classifications apply as previously described (SMM E20.13–16.1–3.1–2.1–2). Descriptions of formwork to edge beams shall include the attached edge of slab (SMM E20.M12), while formwork to recesses, nibs and rebates is measured in metres, stating the number and giving a dimensioned description, and described as extra over the formwork in which they occur on superficial items of formwork (SMM E20.17–19.1.0.1 and E20.M14).

Formwork to wall kickers is measured in metres along the centre line of the wall and is deemed to include both sides (SMM E20.21.0.0.0 and E20.M15). Suspended wall kickers (SMM E20.22) constitute a separate item and occur where a concrete wall rises off a concrete slab, the kicker being cast integrally with the slab and hence, at the time the kicker formwork is erected, there is nothing to support its bottom edge (Code of Procedure: SMM E20.21–22).

*Sequence of measurement of reinforced concrete floors and roofs*

It is important to adopt a logical sequence when measuring reinforced concrete suspended slabs and the student may find the following suggested procedure, coupled with a study of example 6, useful.

(1) *In situ* concrete to slabs, beams and beam casings in m$^3$ (SMM E10.5, 9 & 10).

(2) *In situ* concrete to upstands, deep beams and deep beam casings, if any, in m$^3$, separately classified (SMM E10.9–10.2–3.0.1 and E10.14.0.0.1).

(3) Formwork to soffits of slabs and landings in m$^2$, giving the appropriate slab thickness range and plane (SMM E20.8–9.1–2.1.1–2).

(4) Formwork to beams and beam casings in m$^2$ where of regular shape, stating the number of members (SMM E20.13–14.1–3.1.1–2).

(5) Formwork to sides of upstands in metres where ≤ 1.00 m, high, and classified as plain vertical or giving a dimensioned description (SMM E20.4.1–2.2–4.0).

(6) Adjust formwork to slab where beams occur if not adjusted previously.

(7) Formwork to edges of suspended slabs in metres, if ≤ 1.00 m high, in prescribed height categories, classified as for sides of upstands (SMM E20.3.1–2.2–4.0).

(8) Reinforcement in slabs and beams (bars in kg subsequently reduced to tonnes, keeping each diameter (nominal size) separate, and fabric in m$^2$).

(9) Adjustments if required on supporting brickwork for concrete slabs and beams.

*Measurement of reinforced concrete staircase*

Example 7 illustrates the method of measurement of a reinforced concrete staircase and students are advised to work carefully and methodically through the dimensions, descriptions and explanatory notes to become familiar with the approach adopted for this class of work.

The *in situ* concrete in the steps, waist, string beams and base block are all grouped together in reinforced *in situ* concrete, stating the mix or strength, in staircases (SMM E.10.13). This is followed by the formwork measured in metres to the stairflights, stating the number and widths of stairflights and describing the waists, risers and giving the widths of strings (SMM E20.25.1.1.0). Alternatively, dimensioned diagrams may be used. Formwork to stairflights is deemed to include soffits, risers and strings (SMM E20.C5). Formwork items are however required to spandril ends and the base block, and dimensioned diagrams have been incorporated to help the estimator. Formwork to soffits of landings are dealt with in the same way as soffits of slabs (SMM E20.9.1–2.1–3.1–2).

The concrete and formwork items are followed by the steel reinforcement measured linearly in the first instance, to be subsequently converted to weight. Straight, bent and curved bars are distinguished from those in links, which include stirrups and binders, and each diameter (nominal size) of bar is kept separate (SMM E30.1.1.1–4.0).

The remaining items in the staircase are then taken, including the surface treatment of concrete, where appropriate, granolithic paving to treads and risers and associated work. The metal balustrading is measured under an appropriate heading in metres with a dimensioned or component drawing (SMM L31.2), followed by extra over enumerated items for ornamental ends and the like to handrails. The oak handrail is also a linear item with a separate painting item, followed by enumerated ends where appropriate. Finally the painting of the balustrading is taken in m$^2$ as SMM M60.7.1.1.0.

## PRECAST CONCRETE

Various precast concrete components are measured in *Building Quantities Explained* including a saddleback coping (p. 81), pier caps (p. 81), lintels (pp. 89, 152, 170 and 174), manhole cover slabs (p. 211), channel (p. 217), edging (p. 218) and paving slabs (p. 218). Padstones are included in example 6 and a coping and lintels in example 4 of this book.

The provision of three different units of measurement for precast concrete components in SMM7 permits selection of the most appropriate in each case, such as enumerated pier caps and lintels, linear items for copings and kerbs, and superficial measurement of floors. The descriptions should include the particulars listed in SMM F31.S1–5, namely, kind and quality of materials, mix or strength of concrete, details of reinforcement, surface finishes, bedding and fixing. Formwork or moulds, reinforcement, bedding, fixings, temporary supports, cast-in accessories and pre-tensioning for precast units are all deemed to be included in the measured items (SMM F31.C1).

## PRECAST CONCRETE/COMPOSITE DECKING

Composite construction is a combination within a member (slab, beam or wall) of *in situ* and precast work. SMM E60.P1 prescribes that the location drawings or further drawings accompanying the bill of quantities should show stressing arrangements; full details of anchorages, ducts, sheathing and vents; relative positions of concrete members; size of members; thickness of slabs; and permissible loads.

Composite slabs are measured in m$^2$ stating the thickness, with solid concrete work and filling ends of blocks deemed to be included (SMM E60.1.1.0.0 and E60.C1). Solid concrete work to margins > 500 mm wide are measured as ordinary slabs in accordance with SMM sections E10, E20 and E30. The student should note the slab particulars to be supplied (SMM E60.S1–6), and the normal classifications as to plane apply (sloping≤ and > 15°).

## PRESTRESSED CONCRETE WORK

Concrete members are classed as prestressed concrete work where a stress is artificially induced in the concrete by means of tendons tensioned before working loads are applied. Prestressed work may be either pre-tensioned or post-tensioned. In *pre-tensioning* first the steel is tensioned between abutments and then the concrete is placed in moulds around it. When the concrete has achieved sufficient compressive strength, the steel is released from the abutments, transferring the force to the concrete through the bond that now exists between the steel and the concrete. In *post-tensioning*, the concrete is first cast in the mould and allowed to harden before the prestress is applied. The steel may be placed in position to a predetermined profile and cast into the concrete, bonding being prevented by enclosing the steel in a protective metal sheathing. Alternatively, ducts may be formed in the concrete and the steel passed through after hardening has taken place. When the required concrete strength has been achieved, the steel is stressed against the ends of the unit and anchored off, thus putting the concrete into compression.

Pre-tensioned roof and floor beams are used in buildings, often incorporating infill hollow blocks between the beams. The beams may be rectangular or of inverted T section. Post-tensioning is more versatile and makes more efficient use of the prestressing forces but involves the extra cost of ducts and permanent anchorages. Post-tensioned beams are required for long spans or where they act as continuous members over intermediate supports. For single span members the beams can be cast and post-tensioned on the ground, the cables grouted up and then hoisted into position. An alternative is to use smaller precast sections, either precast reinforced or pre-tensioned, hoisted into position on to falsework, and then to feed the cables through the pre-formed ducts, post-tension and grout. Prestressed concrete can also be used for vertical wall panels and at the Unicorn Hotel and car park, Bristol, the main structural cladding elements consist of attractive 3 m × 3 m precast diamond-shaped units post-tensioned longitudinally and forming an edge beam to support 16 m prestressed floor beams.

Pre-tensioning equipment consists primarily of a temporary grip, normally of a barrel and wedge, which holds the wires or strands during and after tensioning. There is a wide range of post-tensioning systems, which are generally classified by the method adopted to anchor the tendons, and these may be either a threaded nut system or a wedge system. The tendons may be wires, strands or bars.

Drawings of precast concrete units and precast/composite concrete decking are to show stressing arrangements and full details of anchorages, ducts, sheathing and vents (SMM E50/F31/H50/E60.P1a and b).

Precast concrete units are deemed to include pre-tensioning (SMM E50.C1), but the description of the units is to include the kind and quality of pre-tensioning materials, spacing and stresses (SMM E50/F31/H50.S6). In the case of post-tensioned reinforcement for *in situ* concrete, the number of members tensioned is given with a dimensioned description, stating the number of tendons in identical members (SMM E31.1.1.0.0 and E31.M1). The description shall include the number, length, material and size of wires in tendons; ducts, vents and grouting; anchorages and end treatment; stressing sequence, transfer stress and initial stress; and limitation on propping (SMM E31.S1–6).

# DRG. 6      REINFORCED CONCRETE SUSPENDED FLOORS

## REINFORCED CONCRETE SUSPENDED FLOORS

| | | | | |
|---|---|---|---|---|
| | | Structl. stl. framg.<br>(stlwk. to BS 4 Pt. 1, Table 5) | Location drawings, as drg. 6, show the types and sizes of structural members and their positions and details of connections. | |
| | | Framg. fabricatn., beams. comprisg. the follg. items. | It is assumed that the steelwork in this example constitutes structural steel framing (SMM G10) as opposed to isolated structural steel members (SMM G12), as the two beams are jointed together. | |
| | | | The types and grades of materials are inserted in the heading as SMM G10.S1. | |
| | | add beargs.2/150  3·500<br>300<br>3·800 | Add 150 mm bearing, at each end of the beam, to the span to give the total length of the beam. | |
| | 3·80 | Beam 203 × 133 × 30 kg/m U.B.<br>———————— kg | Steel framing, fabrication, beams is measured in tonnes and classified as SMM G10.1.2.0.0. | |
| | | addtnl. len. of beargs.  1·850<br>40<br>1·890 | The lengths of the beams have been inserted in the dimension column and the members are weighted up in the description column. | |
| | 1·89 | Beam, 203 × 133 × 25 kg/m U.B.<br>———————— kg. | Fabrication includes all operations up to and including delivery to site (SMM G10.D1). Fittings are only measured separately where they are of a different type and grade of material (SMM G10.M2), and so the weight of the cleats are added to that of the beams. | |
| 2/ | 0·13 | L cleat 64 × 64 × 6·2 × 5·96 kg/m.<br>———————— kg | | |
| | | End of beams | Items for fabrication measured by weight are deemed to include shop and site black bolts, nuts and washers for structural framing to structural framing connections, (SMM G10.C1), and so the four site black bolts in the beam connection are not separately measured. | |
| | Total wt. of stlwk. | Framg. perm. erectn. on site.<br>(total wt. ___ t). | Permanent erection of framing on site is a separate weighted item as SMM G10.2.2.0.0. | |

6.1

## EXAMPLE 6

| | | | |
|---|---|---|---|
| 3/ | 1 | Precast conc. padstone (1:2:4/20 agg.) 350 × 215 × 150 & beddg. in bwk. in c.m. (1:3). | Padstones are enumerated with a dimensioned description as SMM F31.1.1.0.0, and giving the mix details and form of bedding (SMM F31.S1 and 4). |
| | | R.C. Floor Slabs | The in situ concrete description shall contain the appropriate particulars from SMM E10.S1-5. |
| | 5·85<br>5·75<br>0·15 | In situ conc. slab (1:2:4/19 agg.), thickness ≤ 150, reinfd. | In situ slabs are measured in m³, and stating the thickness range and reinforced classifications as SMM E10.5.1.0.1. |
| | 5·75<br>0·30<br>0·20 | (beam | Concrete beams to suspended slabs are included in with the slabs, except where they are deep beams with a depth/width ratio > 3:1, with the depth measured below the slab (SMM E10.D4a). Although this results in a thicker slab over a small area, it is considered permissible to include the beam concrete in with the slab concrete, without any change of classification. Furthermore, SMM E10.M2 appears to support this approach, as it states that the thickness range given in the descriptions of slabs excludes projections and recesses. |
| | | Fwk.<br>5·850<br>less<br>wall bearg. 215<br>beam 300<br>wall bearg. 112  627<br>5·223<br>5·750<br>less<br>wall bearg. 112<br>102  214<br>5·536 | Indeed it would be unrealistic to measure a small section of slab of the combined thickness in the range 150-450 mm thick.<br>Build up of dimensions of formwork to soffit of slab are inserted in waste to show clearly how they have been computed. |
| | 5·22<br>5·54 | Fwk., soff. of slab, slab thickness ≤ 200, hor., ht. to soff. 1·50 – 3·00 m. | Formwork is classified as SMM E20.8.1.1.2, in accordance with the concrete slab thickness and height to soffit, and stating whether horizontal or sloping ≤ or > 15°. |

6.2

## REINFORCED CONCRETE SUSPENDED FLOORS (Contd.)

| | | | | |
|---|---|---|---|---|
| | | 5.850 | | Formwork to slabs > 200 mm thick is given separately in stages of 100 mm. While formwork to soffits > 1.50 m high is so described stating the height in further stages of 1.50 m. |
| | 2/ | 5.750 | | |
| | | 11.600 | | |
| | | 23.200 | | |
| 23.20 | Fwk. to edges of susp. slab, plain vert., ht. ≤ 250. | | | |
| | | | | It is assumed that the brick skin will be built after the concrete is laid, thus necessitating the provision of formwork to the edges of the floor slab. This is measured in metres giving the appropriate particulars and depth range as SMM E20.3.1.2.0. |
| | gth. of beam | | | |
| | sides 2/200 | 400 | | |
| | soff. | 300 | | |
| | | 700 | | |
| 5.75 | Fwk. to beam, attchd. to slab, reg. rect. shape, ht. to soff. 1.50 – 3.00 m. | | | Formwork to beams attached to slabs is measured in m², stating the number of members, shape and height to soffit (SMM E20.13.1.1.2). Formwork is deemed to include splayed edges (SMM E20.C2). The adjustment of the formwork to the soffit of the slab for the area occupied by the beam has been made previously. |
| 0.70 | | | | |
| | (In 1 nr) | | | |
| | Reinft. | | | |
| | bar lens. | | | |
| | mk 1 | | | |
| | | 3.550 | | Steel bar reinforcement is measured in tonnes keeping each diameter (nominal size) separate and distinguishing between straight, bent and curved bars (SMM E30.1.1.1–3.0). The kind and quality of steel shall be stated (SMM E30.S1). Hooks, tying wire, and spacers and chairs which are at the discretion of the contractor are deemed to be included (SMM E30.C1). Bent bars cover those which are specifically bent to curves to suit the shape of the member in which they are to be placed. Bars are not classified according to location. Links, which include stirrups and binders, form a separate category (SMM E30.1.1.4.0). |
| | add diag. extra | 50 | | |
| | 2 turned ends 2/75 | 150 | | |
| | | 3.750 | | |
| | mk 2a | | | |
| | add | 3.550 | | |
| | 2 turned ends 2/75 | 150 | | |
| | | 3.700 | | |
| | mk 2b | | | |
| | add | 3.200 | | |
| | 2 turned ends 2/75 | 150 | | |
| | | 3.350 | | |
| | mk 3 | | | |
| | | 3.200 | | |
| | add diag. extra | 50 | | |
| | 2 turned ends 2/75 | 150 | | |
| | | 3.400 | | |
| | mk 4 | | | |
| | | 5.750 | | |
| | less cover 2/40 | 80 | | |
| | add | 5.670 | | |
| | turned ends 2/75 | 150 | | |
| | | 5.820 | | |

6.3

| | | | | |
|---|---|---|---|---|
| | mk 5 | | | Horizontal bars and bars sloping ≤ 30° from the horizontal over 12.00 m long are so described, stating the length in further stages of 3.00 m. Whereas with vertical bars and bars sloping > 30° from the horizontal, the commencing length is 6.00 m and thereafter in 3.00 m stages. |
| | | 5.750 | | |
| | less cover 2/40 | 80 | | |
| | add | 5.670 | | |
| | hkd. ends 2/12/20 | 480 | | |
| | | 6.150 | | |
| | mk 6 | | | |
| | len. of mk 5 | 6.150 | | |
| | add diag. extras 2/100 | 200 | | |
| | | 6.350 | | |
| | mk 7 | | | No bar schedule is provided on the drawing in this example, and so the length and number of each bar type has to be calculated. Even if a schedule had been provided it would be necessary to check the information on it. In calculating the lengths, figured dimensions should be used as far as practicable. Adjustments must be made for concrete cover to the reinforcement – usually about 40 mm, and additions for hooked ends (12 times the diameter of the bar), turned ends (often about 75 mm) and the extra for diagonal lengths over straight. |
| | | 300 | 350 | |
| | less cover 2/40 | 80 | 80 | |
| | | 220 | 270 | |
| | | | 220 | |
| | | | 2/490 | |
| | | | 980 | |
| | add turned ends 2/150 | | 300 | |
| | | | 1.280 | |
| 19/ | 3.75 | M.s. bar reinft., 16 φ nom. size, strt. to BS 4449. | (mk.1) | |
| 18/ | 3.70 | | (mk.2a) | Each of the bars are then systematically entered in the dimensions column, taking care to ensure that each type of bar is covered and that the correct timesing figures are inserted. Each different diameter (nominal size) of bar generates a separate item. Bar diameter can be abbreviated to 'dia.' or 'diam.' or shown as φ. The bar type references are inserted in waste for identification purposes. |
| 19/ | 3.35 | | (mk.2b) | |
| 18/ | 3.40 | | (mk.3) | |
| 16/ | 5.82 | Ditto. 6 φ. | (mk.4) | |
| 4/ | 6.15 | Ditto. 20 φ. | (mk.5) | |
| 2/ | 6.35 | | (mk.6) | |
| 35/ | 1.28 | M.s. bar reinft., 6 φ nom. size, links to BS 4449. | (mk.7) | Reinforcement in links is kept separate from that in straight, bent and curved bars (SMM E30.1.1.4.0). |

6.4

## REINFORCED CONCRETE SUSPENDED FLOORS (Contd.)

| | | | | |
|---|---|---|---|---|
| | | mk.8 | | Finally the curved bars around the perforation are taken. The length is calculated from the diameter in each case using the normal approach ($\pi D$) and adding for laps. There is no requirement to state the radius. |
| | | diam. of perf. | 350 | |
| | | add 2/50 | 100 | |
| | | diam. of inner bar | 450 | |
| | | add 2/50 | 100 | |
| | | diam. of outer bar | 550 | |
| 22/7/ | 0.45 | M.s. bar reinft., 10 $\phi$ nom. size, curved to BS4449. | | |
| 22/7/ | 0.55 | | | |
| 2/ | 0.15 | (laps | | |

| | | Conc. Casg. to Stl. Beams | | | The dimensions of the concrete casing to the steel beams are calculated in waste, allowing 50 mm cover to the beams on all sides as shown on the drawing. It is also necessary to adjust the length of the shorter beam at both ends. |
|---|---|---|---|---|---|
| | | | 133 | 203 | |
| | | add cover 2/50 | 100 | 50 | |
| | | | 233 | 253 | |
| | | less | | 1.850 | |
| | | extended bearg. | 40 | | |
| | | ½ width / main beam | 66 | | |
| | | cover to do. | 50 | 156 | |
| | | | | 1.694 | |

| | | | |
|---|---|---|---|
| 3.50 | In situ conc. isoltd. beam casgs. (1:2:4 / 19 agg.). | | The concrete is classified as SMM E10.9.1.0.0 as the beams are below the precast concrete decking. |
| 0.23 | | | |
| 0.25 | | | |
| 1.69 | | | |
| 0.23 | width | 233 | |
| 0.25 | 2/depth 2/253 | 506 | |
| | | 739 | |

| | | |
|---|---|---|
| 3.50 | Fwk. to beam casg. isoltd., reg. rect. shape, ht. to soff. 1.50 – 3.00 m. | Formwork to isolated beam casings is given separately as SMM E20.14.3.1.2, and stating the number of members. It is measured in m² as being of regular shape. |
| 0.74 | | |
| 1.69 | (In 2 nr) | |
| 0.74 | | |

6.5

## CONTRACTOR DESIGNED SUSPENDED FLOOR

| | | |
|---|---|---|
| | | Moulds, reinforcement, bedding, fixings, temporary supports and cast in accessories are deemed to be included in the precast units (SMM E50.C1). |
| 3.50 | Flr. deckg., precast conc. holl. units, thickness: 150, len. 3.40 m, to carry superimposed lds. of flr. & clg. finishgs. of 90 kg/m² & wkg. ld. of 400 kg/m², bedded on bwk. & stl. beam in c.m. (1:4), w. trowelled fin. to tops of slabs & plain fin. to exp. soffs. | Precast concrete floor decking is measured over all bearings in m², with a dimensioned description, stating the length of the floor unit as SMM E50.1. 3.1.0. As the decking is to be contractor designed, it is necessary to insert the requirements as to superimposed loads in the item description. The particulars are to include the form of bedding and surface finishes (SMM E50. S4 and 5). |
| 3.40 | | |

| | | |
|---|---|---|
| | width of slabs | |
| | 2.270 | |
| | add ½ width of beam  62 | |
| | cover to do.   50   112 | |
| | 2.382 | |

| | | |
|---|---|---|
| 2.38 | Ditto., but len. of 1.85 m. | This forms a separate item, because of the different length of floor units. |
| 1.85 | | |

6.6

# DRG. 7 REINFORCED CONCRETE STAIRCASE AND BALCONY
## Sheet 1

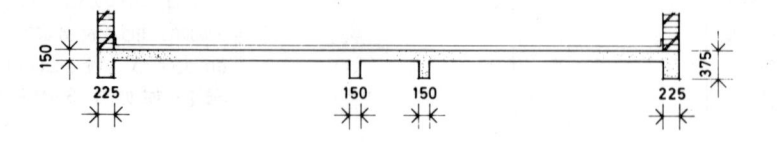

**SECTION A-A** Scale 1:100

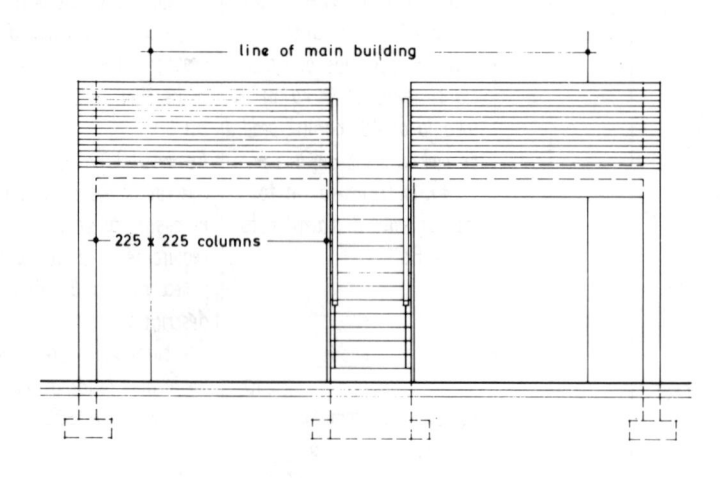

**FRONT ELEVATION** Scale 1:100

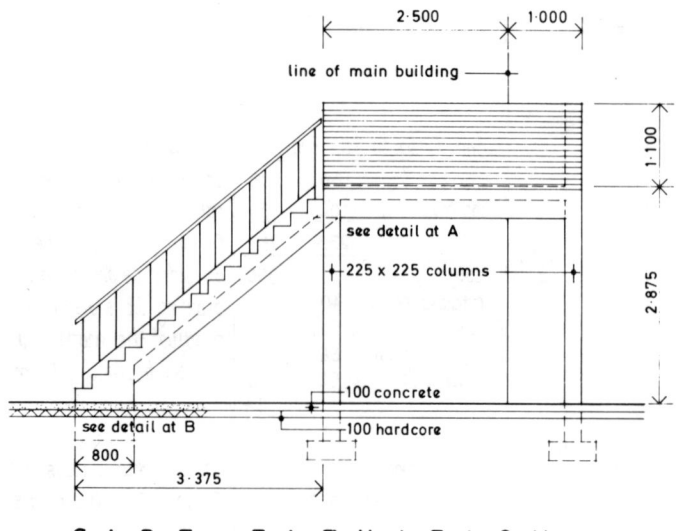

- line of main building
- 225 x 225 columns

**SIDE ELEVATION** Scale 1:100

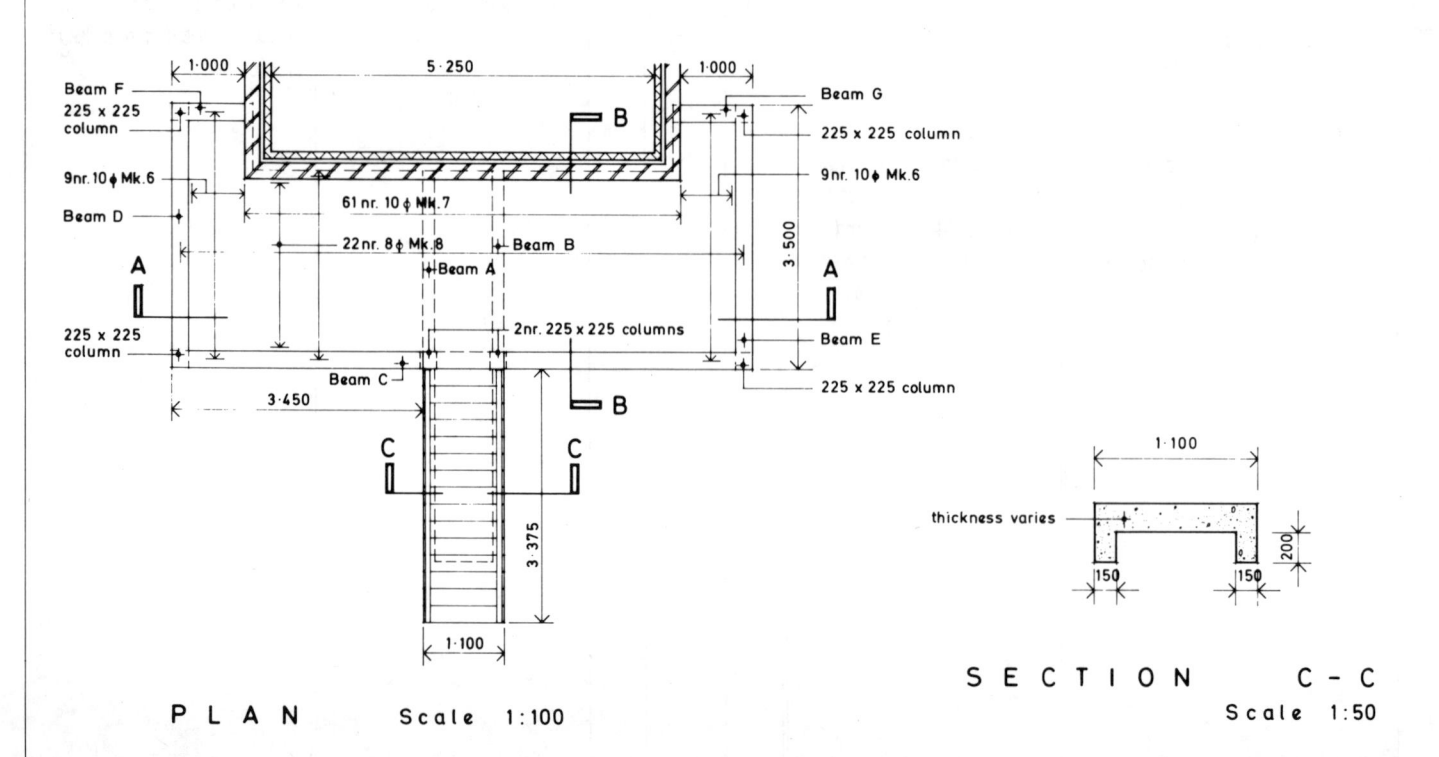

- line of main building
- see detail at A
- 225 x 225 columns
- see detail at B
- 100 concrete
- 100 hardcore

**PLAN** Scale 1:100

- Beam F
- 225 x 225 column
- 9 nr. 10 ⌀ Mk.6
- Beam D
- 225 x 225 column
- Beam G
- 225 x 225 column
- 9 nr. 10 ⌀ Mk.6
- 61 nr. 10 ⌀ Mk.7
- 22 nr. 8 ⌀ Mk.8
- Beam B
- Beam A
- Beam C
- 2 nr. 225 x 225 columns
- Beam E
- 225 x 225 column

**SECTION C-C** Scale 1:50

thickness varies

## DRG. 7  REINFORCED CONCRETE STAIRCASE AND BALCONY
### Sheet 2

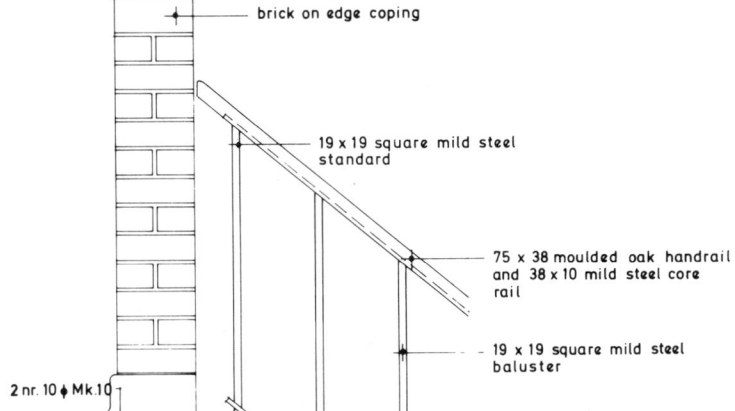

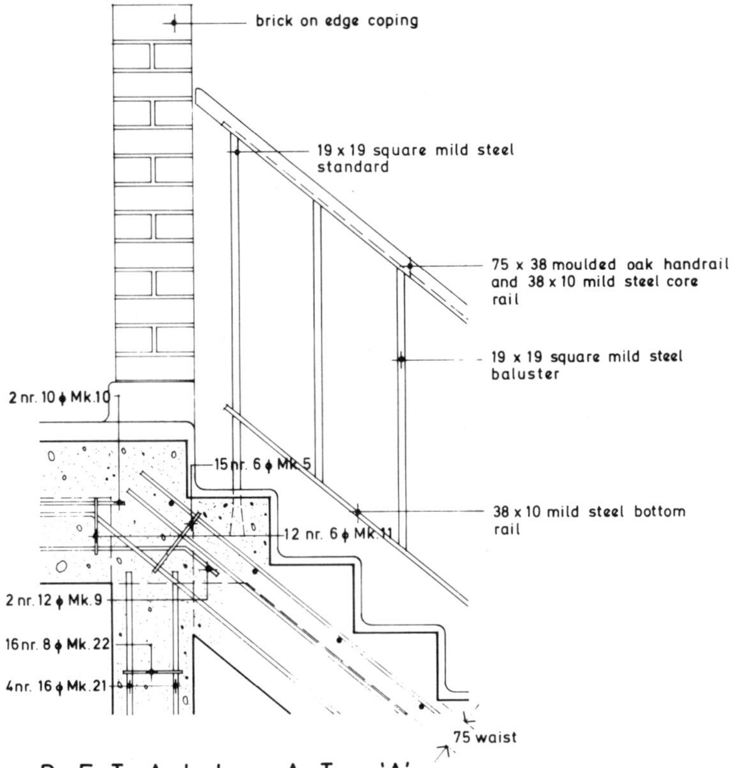

### Specification Notes
All reinforced concrete to be 1:2:4 –19 mm aggregate. All exposed faces to be finished smooth.
Bar reinforcement (as schedule below) to be mild steel to B.S. 4449.
Granolithic paving to be 2 parts cement to 5 parts granite chippings with steel trowelled finish.
Balustrading in mild steel, prepared, welded and ground to smooth finish, primed with calcium plumbate primer and finished with two coats oil paint.

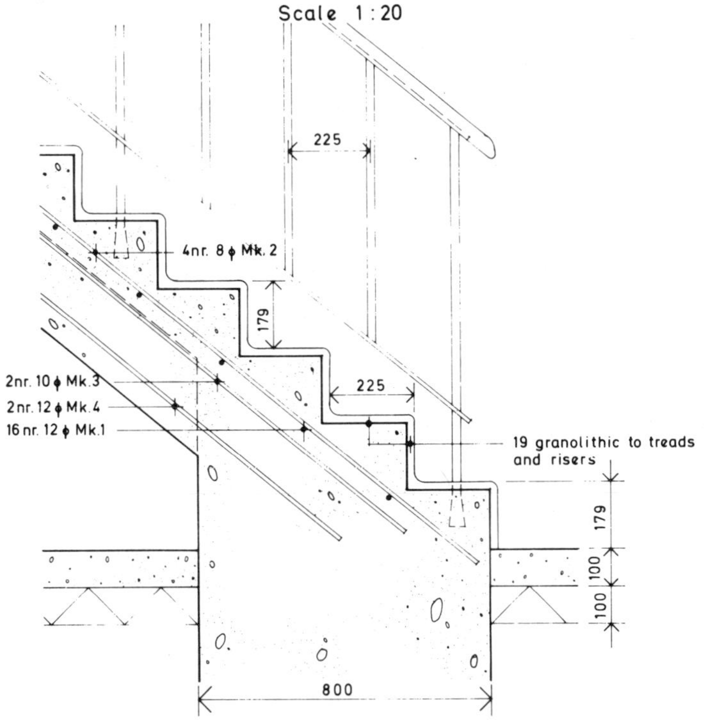

### Reinforcement Schedule

| Member | Mark | Diam. | Nr. | Total Nr. | Girth (mm) | Binder Detail |
|---|---|---|---|---|---|---|
| Stairs | 1 | 12 | 16 | 16 | 1000 | |
| Stairs | 2 | 8 | 4 | 4 | 4550 | |
| String beams to stairs | 3 | 10 | 2 | 4 | 3800 | |
| = = = | 4 | 12 | 2 | 4 | 4250 | |
| Binders | 5 | 6 | 15 | 30 | 800 | |
| Slab | 6 | 10 | 9 | 18 | 3300 | |
| = | 7 | 10 | 61 | 61 | 2450 | |
| = | 8 | 8 | 22 | 22 | 7750 | |
| Beams A and B | 9 | 12 | 2 | 4 | 2200 | |
| = = | 10 | 10 | 2 | 4 | 2000 | |
| Binders | 11 | 6 | 12 | 24 | 1050 | |
| Beam C | 12 | 20 | 3 | 3 | 8175 | |
| = | 13 | 12 | 2 | 2 | 7775 | |
| Binders | 14 | 6 | 40 | 40 | 1010 | |
| Beams D and E | 15 | 20 | 3 | 6 | 3100 | |
| = = | 16 | 12 | 2 | 4 | 2950 | |
| Binders | 17 | 6 | 17 | 34 | 1050 | |
| Beams F and G | 18 | 20 | 3 | 6 | 800 | |
| = = | 19 | 12 | 2 | 4 | 650 | |
| Binders | 20 | 6 | 4 | 8 | 1050 | |
| Columns | 21 | 16 | 4 | 24 | 3000 | |
| Binders | 22 | 8 | 16 | 32 | 700 | |
| = | 23 | 8 | 16 | 64 | 850 | |

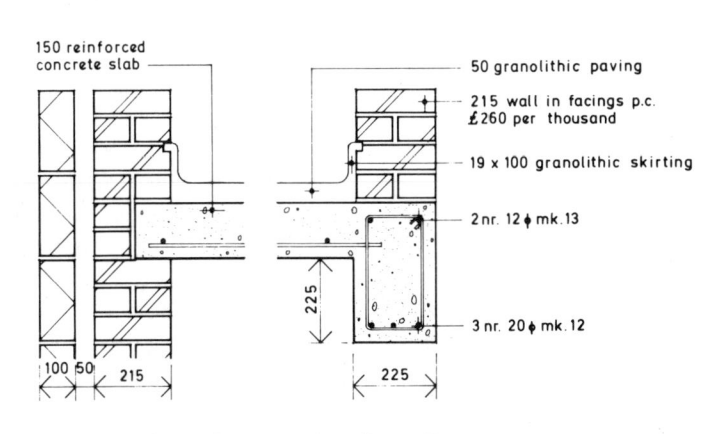

### DETAIL AT 'A'
Scale 1:20

- brick on edge coping
- 19 x 19 square mild steel standard
- 75 x 38 moulded oak handrail and 38 x 10 mild steel core rail
- 19 x 19 square mild steel baluster
- 2 nr. 10 ⌀ Mk.10
- 15 nr. 6 ⌀ Mk.5
- 38 x 10 mild steel bottom rail
- 12 nr. 6 ⌀ Mk.11
- 2 nr. 12 ⌀ Mk.9
- 16 nr. 8 ⌀ Mk.22
- 4 nr. 16 ⌀ Mk.21
- 75 waist

### DETAIL AT 'B'
Scale 1:20

- 225
- 4 nr. 8 ⌀ Mk.2
- 179
- 2 nr. 10 ⌀ Mk.3
- 2 nr. 12 ⌀ Mk.4
- 16 nr. 12 ⌀ Mk.1
- 225
- 19 granolithic to treads and risers
- 179
- 100, 100
- 800

### SECTION B-B
Scale 1:20

- 150 reinforced concrete slab
- 50 granolithic paving
- 215 wall in facings p.c. £260 per thousand
- 19 x 100 granolithic skirting
- 2 nr. 12 ⌀ mk.13
- 3 nr. 20 ⌀ mk.12
- 100, 50, 215
- 225
- 225

## REINFORCED CONCRETE STAIRCASE AND BALCONY

| | | | |
|---|---|---|---|
| | | Reinfd. conc. staircase, cols. & balcony to first flr | The location drawings, equivalent to drg. 7, show the relative positions of concrete members, size of members, thickness of slabs and permissible loads in relation to casting times (SMM E10.P1). It is assumed that the sub-structure has been taken previously and that measurements are to commence from the ground floor slab. The column height is calculated in waste using figured dimensions. |
| | |          Cols. ht. <br> G.F. to F.F. slab <br> (upper surfs.)    2·875 <br> less susp. flr. 150 <br>    beam 225    375 <br>         2·500 | |
| 6/ | 2·50<br>0·23<br>0·23 | In situ conc. cols. (1:2:4/19 agg.) reinfd. <br><br>    4/225 = 900 | Concrete columns are only measured separately when isolated and when their length on plan is ≤ 4 times their thickness as SMM E10.11.0.0.0 and E10.M4. |
| 6/ | 2·50<br>0·90 | Fwk. to cols., isoltd. reg. sq. shape, smth. fin. <br>      (In 6 nr) | Formwork to isolated columns of regular shape is measured in m², stating the column shape and number of members (SMM E20.15.3.1.0). Regular shaped includes, rectangular, circular, hexagonal or other definable regular shape (SMM E20.D10). Passings/intersections of subsidiary beams or other projections are not deducted from areas of formwork (SMM E20.M11). |
| | | Reinft. | |
| 24/ | 3·00 | M.s. bar reinft., 16φ nom. size, strt. to BS 4449. <br>      (mk. 21 | Bar reinforcement is classified as SMM E30.1.1.1.0, with each diameter (nominal size) forming a separate item. The diameter symbol φ has been used in this example as it saves time. There is no requirement to give the location of the reinforcement. The lengths of the binders vary between the different columns. It is vital to check the reinforcement schedule very carefully and to cross through each figure in the total number column as they are entered on |
| 32/ | 0·70 | M.s. bar reinft., <br> 8φ nom. size, <br> links, to BS 4449. | (mk. 22 <br> (under stairs |
| 64/ | 0·85 | | (mk. 23 <br> (corners of <br> (balcony |

7·1

## EXAMPLE 7

| | | | |
|---|---|---|---|
| | |      Balcony <br> Conc. in slab & beams <br>    len. of slab <br>    2/3·450    6·900 <br>        1·100 <br>        8·000 | the dimensions paper. These are classified as links (SMM E30.1.1.4.0). |
| | | len. of edge beam <br>      8·000 <br>      3·500 <br> 2/11·500 <br>    23·000 <br> less corners 4/2/½/225   900 <br>      22·100 | The areas of the suspended slab to the balcony and supporting beams are first calculated in their entirety and adjustments made subsequently for the area occupied for the projection of the main building. |
| | | len. of interm. <br> balcony beams <br>      2·500 <br> less edge beam   225 <br>      2·275 <br> add bearg.    100 <br>      2·375 | The intermediate balcony beams require adjustment of the bearing into the brickwork at one end and the edge beam at the other. |
| | 8·00<br>3·50<br>0·15 | In situ conc. slab <br> (1:2:4/19 agg.), (balcony <br> thickness ≤ 150, (slab <br> reinfd. | The suspended slab falls into the category in SMM E10.5.1.0.1, giving the appropriate thickness range. This item also includes the additional concrete in attached beams, as SMM E10.D4a, but does not result in a variation of the thickness range. |
| | 22·10<br>0·23<br>0·23 | (edge beam | |
| 2/ | 2·38<br>0·15<br>0·23 | (interm. balcony <br> (beams | |
| | |      len. <br>      8·000 <br> less retn. 1·000 <br> bearg. 2/1·112·5   2·225 <br>      5·775 <br>      width <br>      1·000 <br> less bearg.    100 <br>      900 | The adjustments then follow for the area of balcony to be deducted for the projection of the main building and the omission of the rear section of edge beam for the length of the main building, using figured dimensions as far as possible and inserting locational notes in waste. |
| | 5·78<br>0·90<br>0·15 | Ddt ditto. (proj. of main <br> (bldg. into <br> (balcony | |
| | 5·78<br>0·23<br>0·23 | (edge beam <br> (rear len. | The calculations for the area of formwork to the soffit of the balcony are entered in waste, deducting the widths of enclosing beams to which the formwork will be measured separately. |
| | | Fwk. in slab & beams <br>    len. of slab <br>      8·000 <br> less <br> end beams 2/225   450 <br>      7·550 | |

7·2

REINFORCED CONCRETE STAIRCASE AND BALCONY (Contd.)

|  |  |  |  |
|---|---|---|---|
|  |  | less beams (frt.& rear) 2/225 | width 3.500 450 3.050 |

| | | | |
|---|---|---|---|
| | 7.55 3.05 | Fwk., soff. of slab, slab thickness ≤ 200, hor., ht. to soff: 1.50 – 3.00 m, smth. fin. | The formwork to soffits of slabs is measured in m² stating the appropriate slab thickness, plane and height to soffit range as SMM E20.8.1.1.2. Then follows the deduction of formwork for the void occupied by the main building. The width of the rear beam has already been deducted for its entire length. It is assumed that the edges of the slab around the main building will be supported by the new brickwork (shown on Section B-B) and will not therefore require form- work. It is necessary also to deduct the area occupied by the two intermediate downstand beams to which the formwork will be measured separately. The deduction of formwork to the slab soffit area supported by brickwork includes the ends of the two intermediate beams. The length of the formwork to the beams is calculated in waste. No formwork is measured to the ends of the beams supported by brickwork, as it is assumed that the concrete will be poured into pockets in the brickwork. Formwork to the edges of suspended slabs associated with attached beams at slab perimeters is included with the measurement of the form- work to the beams (SMM E20. M12). As the formwork is of regular shape, it is measured in m², stating the shape and height to soffit range (SMM E20. 13.1.1.2). |

Proj. of main bldg.
8.000
less 2/1.000    2.000
6.000
1.000
less rear beam  225
775

| | | | |
|---|---|---|---|
| | 6.00 0.78 | Ddt. last | (slab |
| 2/ | 2.28 0.15 | | (2 interm. beams |

Edge downstd. beams
len.
22.100
less proj. of main bldg.  6.000
16.100
ath.
375
225
225
825

Interm. downstd. beams
len.
2.375
less bearg.  100
2.275
ath.
2/225  450
150
600

| | | | |
|---|---|---|---|
| | 16.10 0.83 | Fwk. to beams attchd. to slab, reg. rect. shape, ht. to soff. 1.50 – 3.00 m, smth. fin. (In 9 nr) | (edge beams |
| 2/ | 2.28 0.60 | | (inter. beams |

7.3

formwork to edge downstand beam

Where beams pass subsidiary beams or other projections, no deductions are made to formwork, but such intersections are deemed to constitute the commencement of an additional member (SMM E20. M11). Formwork to ends of members is deemed to be included with the items (SMM E20. C3). The formwork to the two intermediate downstand beams is taken next, and these are of straightforward regular shape. The reinforcement in suspended slabs can be combined with that in the attached beams, where of similar nominal size, as SMM 7 contains no locational requirements for reinforcement.
Bar reinforcement will be measured in the bill of quantities in tonnes, and the weights of the bars of the prescribed diameters (nominal size) can be obtained from BS 4449, steelwork tables and various other reference books. Hence the weights of the various bars used are as follows:

| nominal size(φ) mm | weight kg/m |
|---|---|
| 6 | 0.222 |
| 8 | 0.395 |
| 10 | 0.616 |
| 12 | 0.888 |
| 16 | 1.597 |
| 20 | 2.466 |

The locations of the bars and their mark numbers are given in waste for purposes of identification.

Reinft. in slab & beams

| | | | |
|---|---|---|---|
| 18/ | 3.30 | M.s. bar reinft., 10φ nom. size, strt. to BS 4449. | (slab mk.6 |
| 61/ | 2.45 | | (slab mk.7 |
| 4/ | 2.00 | | (beams A.& B mk.10 |
| 22/ | 7.75 | Ditto. 8φ. | (slab mk.8 |
| 4/ | 2.20 | Ditto. 12φ. | (beams A.& B mk.9 |
| 2/ | 7.78 | | (beam C mk.13 |
| 4/ | 2.95 | | (beams D.& E mk.16 |
| 4/ | 0.65 | | (beams F.& G mk.19 |
| 3/ | 8.18 | Ditto. 20φ. | (beam C mk.12 |
| 6/ | 3.10 | | (beams D.& E mk.15 |
| 6/ | 0.80 | | (beams F.& G mk.18 |

7.4

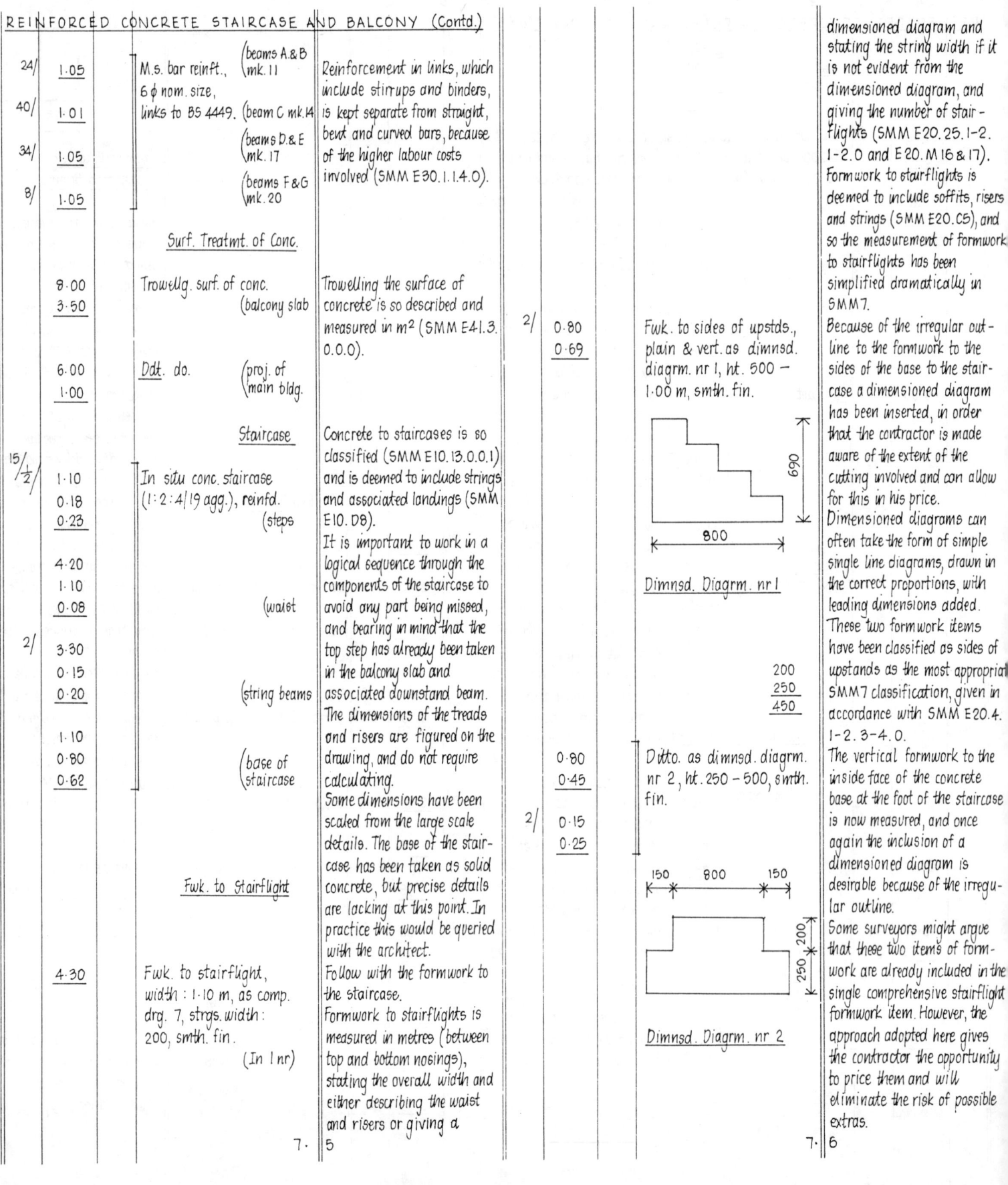

## REINFORCED CONCRETE STAIRCASE AND BALCONY (Contd.)

| | | | | | | | |
|---|---|---|---|---|---|---|---|
| 24/ | 1.05 | M.s. bar reinft., 6 ø nom. size, links to BS 4449. | (beams A & B mk.11 | Reinforcement in links, which include stirrups and binders, is kept separate from straight, bent and curved bars, because of the higher labour costs involved (SMM E30.1.1.4.0). | | | |
| 40/ | 1.01 | | (beam C mk.14 | | | | |
| 34/ | 1.05 | | (beams D & E mk.17 | | | | |
| 8/ | 1.05 | | (beams F & G mk.20 | | | | |

Surf. Treatmt. of Conc.

| | | | | | | | |
|---|---|---|---|---|---|---|---|
| | 8.00 | Trowellg. surf. of conc. | | Trowelling the surface of concrete is so described and measured in m² (SMM E41.3.0.0.0). | 2/ | 0.80 0.69 | Fwk. to sides of upstds., plain & vert. as dimnsd. diagrm. nr 1, ht. 500 – 1.00 m, smth. fin. |
| | 3.50 | | (balcony slab | | | | |
| | 6.00 | Ddt. do. | (proj. of main bldg. | | | | |
| | 1.00 | | | | | | |

Staircase

Concrete to staircases is so classified (SMM E10.13.0.0.1) and is deemed to include strings and associated landings (SMM E10.D8).

| | | | | |
|---|---|---|---|---|
| 15/½/ | 1.10 | In situ conc. staircase (1:2:4/19 agg.), reinfd. | | |
| | 0.18 | | | |
| | 0.23 | | (steps | |
| | 4.20 | | | |
| | 1.10 | | | |
| | 0.08 | | (waist | |
| 2/ | 3.30 | | | |
| | 0.15 | | | |
| | 0.20 | | (string beams | |
| | 1.10 | | | |
| | 0.80 | | | |
| | 0.62 | | (base of staircase | |

It is important to work in a logical sequence through the components of the staircase to avoid any part being missed, and bearing in mind that the top step has already been taken in the balcony slab and associated downstand beam. The dimensions of the treads and risers are figured on the drawing, and do not require calculating.

Some dimensions have been scaled from the large scale details. The base of the staircase has been taken as solid concrete, but precise details are lacking at this point. In practice this would be queried with the architect.

Follow with the formwork to the staircase.

Formwork to stairflights is measured in metres (between top and bottom nosings), stating the overall width and either describing the waist and risers or giving a

Fwk. to Stairflight

| | | |
|---|---|---|
| 4.30 | Fwk. to stairflight, width : 1.10 m, as comp. drg. 7, strgs. width : 200, smth. fin. (In 1 nr) | |

|   |   |   |
|---|---|---|
| 2/ | 0.80 0.45 | Ditto as dimnsd. diagrm. nr 2, ht. 250 – 500, smth. fin. |
| 2/ | 0.15 0.25 | |

Dimnsd. Diagrm. nr 1

```
     800
  ___
 |   |___
 |       |___       690
 |_____|
```

|   |
|---|
| 200 |
| 250 |
| 450 |

Dimnsd. Diagrm. nr 2

```
 150  800  150
  _____
 |  _____  |   200
 |_|       |_|   250
```

dimensioned diagram and stating the string width if it is not evident from the dimensioned diagram, and giving the number of stair-flights (SMM E20.25.1-2.1-2.0 and E20.M16 & 17). Formwork to stairflights is deemed to include soffits, risers and strings (SMM E20.C5), and so the measurement of formwork to stairflights has been simplified dramatically in SMM7.

Because of the irregular outline to the formwork to the sides of the base to the staircase a dimensioned diagram has been inserted, in order that the contractor is made aware of the extent of the cutting involved and can allow for this in his price. Dimensioned diagrams can often take the form of simple single line diagrams, drawn in the correct proportions, with leading dimensions added. These two formwork items have been classified as sides of upstands as the most appropriate SMM7 classification, given in accordance with SMM E20.4.1-2.3-4.0.

The vertical formwork to the inside face of the concrete base at the foot of the staircase is now measured, and once again the inclusion of a dimensioned diagram is desirable because of the irregular outline.

Some surveyors might argue that these two items of formwork are already included in the single comprehensive stairflight formwork item. However, the approach adopted here gives the contractor the opportunity to price them and will eliminate the risk of possible extras.

7.5      7.6

## REINFORCED CONCRETE STAIRCASE AND BALCONY (Contd.)

### Reinft. to Staircase

| | | | |
|---|---|---|---|
| 16/ | 1.00 | M.S. bar reinft., 12φ nom. size, strt. to BS 4449. | (stairs mk.1) (stairs mk.4) |
| 4/ | 4.25 | | |
| 4/ | 4.55 | Ditto. 8 φ. | (stairs mk.2) |
| 4/ | 3.80 | Ditto. 10 φ. | (string beams mk.3) |
| 30/ | 0.80 | M.S. bar reinft., 6φ nom. size, links, to BS 4449. | (binders mk.5) |

The bar reinforcement is measured in a logical sequence separating the bars of different diameters and also the straight bars from those forming links, which include stirrups and binders (SMM E30.1.1.1 & 4.0).

The reinforcement entries are checked and the bars are suitably identified in waste.

### Surf. Treatment of Conc.

| | | | |
|---|---|---|---|
| 15/ | 1.10 | Trowellg. surf. of conc. | (treads) |
| | 0.23 | | |
| 16/ | 1.10 | Hackg. surf. of conc. | (risers) |
| | 0.18 | | |

Trowelling concrete to treads is measured in m² (SMM E41.3.0.0.0).

Hacking surfaces of concrete is given in m² (SMM E41.4.0.0.0).

### Granolithic Paving

**T.O. List**
1. Grano. to balcony
2. Sktg. to balcony
3. R.o.j. of bwk. to form key
4. Grano. to treads & risers

To demonstrate some of the steps taken in practice a query list and take off list are included for this section of the work. The take off list provides the quantity surveyor with the opportunity to look at the particular category of work in its entirety, to list the components in a logical sequence, which then provides a check list as the detailed measurement proceeds.

**Query List**

Q
1. Is carborundum powder to be sprinkled on surf. of grano. pavg.  6/7/89

A
No carborundum reqd. 7/7/89

```
               len.
              8.000
less bk.walls 2/215  430
              7.570
              width
              3.500
less bk.walls 2/215  430
              3.070
```
(want - 6.000 × 1.000)

The dimensions of the area of granolithic paving to the balcony are entered in waste, making adjustments for the enclosing brick wall.

7/7

### External Work

| | | | |
|---|---|---|---|
| 7.57 | | Grano. flr., lev. & to falls ≤ 15° from hor., thickness: 50, 2 cts. ea. of 2pts. P. ct. to 5pts. granite chippgs. graded 6 to dust, w. stl. trowelled fin. on conc. | (balcony) (opg. thro. wall from staircase) |
| 3.07 | | | |
| 1.10 | | | |
| 0.22 | | | |
| 6.00 | | Ddt. ditto. | (proj. of main bldg.) |
| 1.00 | | | |

```
                   Sktg.
                   7.570
                 2/3.070
                  10.640
                  21.280
add retns. to
main bldg. 2/800   1.600
                  22.880
less
opg. to staircase  1.100
                  21.780
add
retns.to opg.2/215  430
                  22.210
```

| | | | |
|---|---|---|---|
| 22.21 | | Grano. sktg., ht: 100, thickness 19, 2:3 mix, stl. trowel fin. on bwk. | |
| | | & | |
| | | R.o.j. of bwk. to form key for grano. sktg. | |
| | | × 0.10 | |

### The follg. in ext. staircase area

| | | |
|---|---|---|
| 1.10 | Grano. treads, width: 225, thickness: 19, mix: 2:5, w. stl. trowld. fin. on conc. | |

Granolithic flooring is deemed to be internal, unless described as external (SMM M10.D1), and this is covered in the heading. The paving to the balcony is classified as to floors, being the nearest appropriate equivalent, and stating the plane, thickness and number of coats as SMM M10.5.1.1.0. The description of the granolithic is to include the kind, quality, composition and mix of materials, method of application and nature of base as SMM M10.S1-6.

Build up of girth of granolithic skirting to balcony, using the dimensions previously calculated for the surface area, with the necessary adjustments.

Granolithic skirtings are measured in metres stating the height, width or girth and thickness as SMM M10.13.0.1.0. Skirtings are deemed to include fair edges, rounded edges, beaded edges, coved junctions, ends, angles and ramps (SMM M10.C10).

Raking out joints of brickwork to form a key for the skirting are measured in m² stating the type and purpose of the work and type of wall (SMM F10.26.1.1.0).

In this instance the alternative method of entering a superficial item following an equivalent linear item has been adopted for the guidance of the student.

Work in staircase areas is to be given separately (SMM M10.M3), and hence the insertion of the general heading. Granolithic treads are measured in metres, stating the width and thickness (SMM M10.7.0.1.0) and the appropriate particulars from SMM M10.S1-6.

7/8

## REINFORCED CONCRETE STAIRCASE AND BALCONY (Contd.)

| | | | | |
|---|---|---|---|---|
| 16/ | 1.10 | Grano. risers, plain, ht: 179, thickness: 19, mix: 2:5, w. stl. trowld. fin. on conc. | | Risers are measured in metres, stating whether plain or undercut, and height and thickness as M10.8.1.1.0. Alternatively, they could be described as 19 thick × 179 high as in the SMM 7 Library of Standard Descriptions. Treads and risers are deemed to include fair edges, internal and external angles, and intersections ≤ 10 mm radius (SMM M10.C6). It is assumed that the angles are 10 mm radius and so do not require including in the description. The work is deemed to include working around obstructions, pipes and the like, into recesses and shaped inserts (SMM M10.C1b), and so no items are required for working the granolithic around the mild steel standards to the staircase. The balustrading is measured in metres as isolated balustrades with a dimensioned description or component drawing as SMM L31.2.1-2.0.0. Isolated balustrades are those which do not form an integral part of a staircase unit (SMM L31.D2). Fixings, such as jointing the core rail to the handrail, are deemed to be included (SMM L31.C1e), but forming mortices in concrete for feet of standards and grouting are separately enumerated. The handrail is classified as an associated handrail as it is of a different material from the balustrade with which it is associated (SMM L30.D3). Associated handrails are measured in metres, giving a dimensioned description as SMM L30.3.1.0.0. |

Balustrdg.

2/ 4.00  Isoltd. balustrade, 1080 hi. o/a, vert., galvd. m.s. 38 × 10 rails & 19 × 19 balusters & stands. as comp. drgs. A & B.

Timber handrail

```
                    4.000
add ends 2/150       300
                    4.300
```

2/ 4.30  Asstd. hdrl. 75 × 38 wrot. mo. oak w. stopped grve., fixg. w. stainless stl. screws & selected & protected for subsequent treatment.

7.9

```
              gth
               75
            2/  38
               113
               226
less core rail  38
               188
```

2/ 4.30  Clear finishg. gen. isoltd. surfs., wd., gth. ≤ 300, prep. & 2 cts. of linseed oil, ext.

2/ 4.00
   1.08  Paintg. met. railgs., plain open type of 38 × 10 rls. & 19 × 19 bals., as comp. drg. 7, detail A, wire brush, prime w. cal. plumbate & apply 2 cts. of oil paint, ext. (one side only mesd.), gth. > 300.

Sundries

2/ 5  Fwk. to in situ conc. mors., gth ≤ 500, depth ≤ 250, rect.

&

Fillg. mors. in in situ conc. w. grout.

The description of the timber shall include the kind and quality and whether sawn or wrought and if selected and protected for subsequent treatment (SMM L30.S1-4). Also the method of fixing where not at the discretion of the contractor (SMM L30.S8). Painting/clear finishing to rails are classified as to general isolated surfaces, girth ≤ 300 mm, as SMM M60.1.0.2.0, and described as external (SMM M60.D1). The rounded ends to the handrail are deemed to be plain ends as distinct from ornamental ends (SMM L30.C3 and L30.4.4.0.0) and have not therefore been measured.
Painting of these balustrades is classified as to railings, plain open type, and measured one side only in m² as girth > 300 mm (SMM M60.7.1.1.0). The particulars of the paint shall include the kind and quality of paint, nature of base, preparatory work, and number of priming and other coats (SMM M60.S1-6). They are further classified according to the size of their individual members (SMM M60.M10). With their straightforward outline they cannot properly be classified as ornamental type railings and measured both sides as SMM M60.7.3.0.0 and M60.M12.
Enumerated item with description as SMM E20.26.1.1.1. to cover the formwork needed to form the mortices to accommodate the feet of the standards to the staircase. Best covered by a separate enumerated item as SMM E10.17.1.0.0.

7.10

# FOR READER'S NOTES

# DRG. 8 — PRESTRESSED PRECAST CONCRETE BRIDGE

## ABUTMENT DETAILS  Scale 1:50

- 2·100
- Freyssinet anchorages and recesses
- concrete mix 30 N/mm² (20 aggregate)
- 12 φ bars at 150 ccs. both ways
- upstand beam
- 20 asphalt
- 200
- 600
- 100 thick joints in in-situ concrete 30 N/mm² (10 aggregate)
- cement mortar (1:3)
- prestressing cables in corrugated steel sheathing to be post tensioned. Ducts grouted with colloidal cement. Freyssinet cone anchorages
- concrete mix 30 N/mm² (20 aggregate)
- concrete mix 20 N/mm² blinding
- 1·000
- 1·600
- 100
- 2·600 × 2·400 wide

## SECTION A-A  Scale 1:50

- Cables to be stressed in sequence of row 1 followed by row 2 in section A-A
- 100 in-situ concrete 30 N/mm²
- prestressed precast concrete units 2·10 m long
- 12 φ bars at 200 ccs
- 200
- 6 φ stirrups at 300 centres
- formwork to exposed concrete to be hardboard lined
- 22 nr. 10 φ bars
- row 1
- row 2
- 400
- 6 φ stirrups at 200 ccs
- 4 prestressing cables in each beam
- 600 / 1·200 / 600
- 2·400
- 200 / 50 / 150 / 400 / 600

## ELEVATION  Scale 1:200

- 28·850
- 29·050
- 28·850
- new road surface
- 20·700
- 20·000
- 20·000
- existing ground level ✦ 28·220
- finished level ✦ 28·220

## PLAN  Scale 1:200

- 28·700  27·476  A  200 × 200 upstand beam  27·349  29·000
- 28·850  29·050  28·850
- 28·550  27·326  A  100 thick in-situ concrete joint  27·274  28·925
- 2·200 × 8 = 17·600
- 19·600

## DETAILS OF FREYSSINET ANCHORAGE   N.T.S (not to scale)

- 50 φ duct
- helical core inside duct
- reinforced concrete anchorage block 120 φ cast into end precast unit
- 12 nr. 6 φ wires of cable on helical core stressed with Freyssinet jack
- male anchorage cone
- grout hole
- grooves for wire

# PRESTRESSED PRECAST CONCRETE BRIDGE

The follg. in prestressed precast concrete bridge 19.60 m long × 2.40 m wide, formed in 8 nr segments to the outline shown on Drg. 8. The bridge is to be post tensioned using the Freyssinet system. 4 nr prestressing cables threaded thro. corrugated stl. sheathg. are to be used in each downstand beam, ea. consisting of 12 nr. 6 φ steel wires to BS 2691, on a helical core stressed with a reinforced concrete anchorage block and Freyssinet jacket, tensioning all 12 wires simultaneously, & the cables in the sequence shown on the drg. The prestressing force is to be 40 t/cable & the strength of the conc. is to be 35 N/mm² at transfer of stress. Ends of wires are to be cut off after completion of the prestressing. Anchorages are to consist of reinforced conc. blocks, 120 lg., cast into end precast segments ea. w. a. male grooved anchorage cone. Freyssinet jacks are to be used tensioning both ends of tendons.

With prestressed precast concrete work of this kind, one approach is to start with an appropriate heading, stating the method of tensioning, and other relevant particulars and referring to the location or component drawing. The location drawings or further drawings accompanying the bill of quantities should show (a) details of precast concrete members showing stressing arrangements; (b) full details of anchorages, ducts, sheathing and vents; (c) the relevant positions of concrete members; (d) size of members; (e) thickness of slabs; and (f) permissible loads, as drawing 8 (SMM E50.P1).

The concrete to the abutments has been omitted from this example as the method of measurement follows the principles previously described. The term 'tendon' is applied to a stretched element used in a concrete member to impart prestress to the concrete. Tendons may consist of individual hard drawn wires, bars or strands. There are a variety of proprietary prestressing systems.

As the tendons are tensioned after the concrete is cast (post-tensioned), the particulars listed in SMM E50.P1 must be incorporated in the appropriate drawings and may be incorporated in a heading in the bill of quantities to assist the estimator.

Calculation of lengths of stirrups:

downstand beams      520
                  2/ 300
                     820
                   1·640
       add laps 2/50  100
                   1·740

Nr of bars

200 ) 2·100
     10+1 = 11

300 ) 2·100
      7+1 = 8

Len. of hkd. bars

                     2·400
less cover 2/40         80
                     2·320
add bends 2/12/12      288
                     2·608

8·1

## EXAMPLE 8

### Precast Units

|6|

Precast reinfd. conc. bridge units 2·10 m lg., 2·40 m wide & 800 mm hi. o/a., as detailed on Drg. 8 w. 2 nr downstd. beams & 2 nr upstd. beams, all in conc. w. a strength of 30 N/mm² (20 agg.) & reinfd. w. 22 nr 10 φ strt. m.s. bars 2·10 m lg., 11 nr 12 φ strt. m.s. bars 2·61 m lg., 22 nr 6 φ m.s. stirrups 1·74 m lg. & 16 nr 6 φ m.s. stirrups 720 lg. all to BS 4449, all exposed surfs. to be fin. smth. w. hdbd. lined fwk. The units are assembled and post-tensioned after erection & are self-supporting only after tensioning, to a longtdnl. gradient of 1 in 30. Post tensng. arrangements as described in the main headg. are to be included.

upstand beams      215
                   120
                 2/ 335
                   670
   add laps 2/25   50
                   720

These precast concrete units are best enumerated with a dimensioned description and giving the reinforcement details as SMM E50.1.1.0.1, and also the particulars listed in E50.S1-6, comprising mix or strength requirements of concrete; kind, quality and size of materials; surface finishes; bedding and fixing. The lengths of reinforcing bars and stirrups have been given in the description, because of the large quantity involved. Straight bars are deemed to include hooked ends (SMM E30.C1). The unit descriptions are to include the post tensioning arrangements as only the pre-tensioning is deemed to be included (SMM E50.C1). On the other hand, post tensioned in situ concrete members are enumerated, stating the number of tendons in identical members (SMM E31.1.1.0.0 and E31.M1), and giving full particulars of the constructional details (SMM E31.S1-5).

It has been verified from the designer that the units will be self-supporting only after tensioning.

8·2

| | | | | | | | |
|---|---|---|---|---|---|---|---|
| | | | PRESTRESSED PRECAST CONCRETE BRIDGE (Contd.) | | | | |
| | 2 | | Ditto. but formed solid for 1.60 m of length as shown in detail on component drg. RCB 2X & the m.s. bar reinft. detailed on the drg., inc. a mesh of 12φ bars at 150 ccs. in both directns. This unit will be self-supportg. both before & after erectn, & bedded in c.m. (1:3) on in situ conc. abutment (m/s), incldg. constn. jt. of asp. 20 th. for full profile of o.e. of unit. Post tensng. included a.b. | The two end units are of different outline with a different arrangement of reinforcement. Because of the complexity reference will need to be made to a detailed component drawing to be supplied by the designer. Alternatively a component drawing could be incorporated in the Bill. These units will be self-supporting from the abutments. As a further alternative a full description of the unit and its reinforcement could be given but it would prove difficult. Construction joints can be enumerated, with a dimensioned description as SMM E50.3.2.1.0 or, alternatively, as adopted in this example, enumerated joints may be given in the description of the precast items in which they occur as SMM E50.M4. | 17.60<br>2.00<br>0.10 | In situ conc. bed, strength: 30 N/mm$^2$ thickness ≤ 150, slopg. ≤ 15°. | Beds of in situ concrete are measured in m$^3$ and giving the thickness classification as SMM E10.4.1.0.3. |
| | | | | | 17.60<br>2.00 | Trowellg. surf. of conc., slopg. | Trowelling the surface of concrete is described and measured in m$^2$ (SMM E41.3.0.0.1). |
| | | | | | | | Another alternative design would be to use pre-tensioned beams spanning the total length (17.60m). |
| | 7 | | Constn. jt. of in situ conc., strength: 30N/mm$^2$ (10 agg.) 100 th., between bridge units after assembly, to the profile shown on Drg. 8. | The thicker construction joints between the units will be more complicated as they will be formed after each adjoining unit is hoisted into position.<br>Formed joints in in situ concrete are deemed to include formwork (SMM E40.C1).<br>Temporary supports or falsework are deemed to be included (SMM E50.C1). | | | |
| | | | <u>In situ conc.</u><br>less        2.400<br>upstd. beams 2/200  <u>400</u><br>                           <u>2.000</u> | Adjustment of width for upstand beams. | | | |
| | | | 8.3 | | | | 8.4 |

# 5 STRUCTURAL STEELWORK

## GENERAL INTRODUCTION

In practice, steelwork may be covered by prime cost sums in bills of quantities, and firms of steel fabricators and erectors prepare their estimates direct from the structural engineer's drawings and specification. It is, however, becoming increasingly common for quantity surveyors to measure this class of work and examination questions are regularly set on this subject.

Two connected steel beams were incorporated in example 6 (Reinforced Concrete Suspended Floors), while the example in this chapter encompasses a framework of columns, beams and grillages. Although the example is not over extensive in its scope, it does cover the main components of structural steel framing and illustrates a methodical approach to the measurement of this class of work.

## PRINCIPLES OF MEASUREMENT OF STRUCTURAL STEEL FRAMING

The information required on location drawings or on further drawings which accompany the bill of quantities under SMM G10.P1 usually comprises the engineer's fabrication or framing drawings. A general description of the work may be given in a heading to the structural framing, where the location drawings are insufficiently precise, and this will indicate briefly the nature of the work and its location as an introduction to the more detailed information given elsewhere. This approach has been adopted in example 9.

Ideally, the drawings should conform to SMM G10.P1 and give information on (a) the position of work in relation to other parts of the work and of the proposed buildings; (b) the types and sizes of structural members and their positions in relation to each other; and (c) details of connections or of the reactions, moments and axial loads at connection points. Connections include fittings used to form a joint and thus enable members to be united by bolting, welding or riveting in the shop or on the site (Code of Procedure G10.P1).

The fabrication of structural steel framing and isolated structural metal members is kept separate from the erection (SMM G10/12, 1 and 2), although both are measured in tonnes, but are entered in the bill to two places of decimals (nearest 10 kg), in accordance with SMM General Rules 3.3. The erection item is a single omnibus item, giving the total weight of steelwork to be erected. Trial and permanent erections are kept separate, and erection includes all operations subsequent to fabrication (SMM G10.D5).

SMM G10, clauses S1–3 list the various particulars of the steel that are to be supplied, comprising types and grades of materials and details of welding tests, X-rays and performance tests. The Code of Procedure amplifies this information by requiring details of the location of the tests and other supporting test requirements.

*Fabrication of framing*
Fabrication of structural steel framing is classified according to the type of member, such as columns, beams, bracings, purlins and cladding rails, grillages, and trusses and built up girders (SMM G10.1.1–8). The mass of framing includes all components and fittings, except fittings of a different type and grade of material (SMM G10.M1). Items for fabrication measured by weight are deemed to include shop and site black bolts, nuts and washers for structural framing to structural framing connections, which do not therefore require specific mention (SMM G10.C1). Fabricated components not measured by weight consist of holding down bolts or assemblies, and special bolts and fasteners (enumerated), permanent formwork in m$^2$, and cold rolled purlins and cladding rails which are measured in metres.

Structural steel members are usually entered on dimensions paper as linear items, stating the depth and breadth and the weight in kg per metre, to be subsequently reduced to weight in kg. Members of similar dimensions can have varying weights, such as universal beam sections 203 × 133 with weights of 25 or 30 kg per metre. The necessary data for structural steel members can be obtained from structural steelwork tables. The overall lengths of the members are measured with no deductions for splayed cuts or mitred ends or for the mass of metal to form notches and holes each < 0.10 m$^2$ in area measured in the plane (SMM G10.M3). All members other than straight and of uniform section, such as castellated, tapered, curved and cambered, shall each be so described. When computing the weight, no allowance is made for the mass of weld fillets, black bolts, nuts, washers, rivets and protective coatings (SMM G10.M4).

*Fittings, connections and special bolts*
Fittings to structural steel framing encompass caps, bases, gussets, end plates, splice plates, cleats, brackets, stiffeners, distance pieces, separators and the like. They are weighted up with the members to which they are attached, unless they are of a different type and grade of material, when they will be measured separately by weight in accordance with SMM G10.1.10.0.0 and G10.M2. Example 9 shows one approach to the measurement of structural steel members and their associated fittings.

SMM7 reduces drastically the number of items to be separately enumerated to two categories. These are (1) holding down bolts or assemblies, giving the requisite details, such as diameter and length for holding down bolts, and probably drawings for the assemblies (SMM G10.1.11.1.0); and (2) special bolts and fasteners, stating the type and diameter (SMM G10.1.12.1.0). Special bolts and fasteners are defined as those other than black bolts and holding down bolts and assemblies (SMM G10.D4), and a good example would be high strength friction grip bolts.

*Surface preparation and treatment*
Surface preparation of steelwork is measured in m$^2$, stating the type of preparation, such as blast cleaning, pickling, wire brushing, flame cleaning and other suitable processes (SMM G10.7.1–5.0.0). Surface treatments are dealt with similarly, giving details of the treatment, such as galvanising, sprayed metal coating, protective painting and other appropriate applications (SMM G10.8.1–4.0.0). Details of the timing of both preparation and treatment are to be given in accordance with SMM G10.S4. The SMM7 Library of Standard Descriptions categorises the timing as 'at works' and 'on site, prior to erection and after erection'.

## MEASUREMENT PROCEDURE FOR STRUCTURAL STEEL FRAMING

As with all measurement work, it is advisable to adopt a logical sequence to avoid the omission of items and to secure higher marks in the examination. The following check list should prove helpful to the student.

*General procedure*
(1) The following steel members are entered as linear items on the dimensions paper
    (a) columns
    (b) beams
    (c) bracings
    (d) purlins and cladding rails
    (e) grillages (component members)
    (f) overhead crane rails
    (g) trestles, towers and built up columns (component members)
    (h) trusses and built up girders (component members)
    (i) wires, cables, rods and bars
    (j) angle sections in cleats and the like.
(2) Plates of all shapes and thicknesses are entered as superficial items on the dimensions paper, averaging the dimensions of irregularly shaped plates.
(3) The weighting up process usually takes place at the bill production stage when the total measured length or area of each member or component is multiplied by the weight per m or m$^2$ as appropriate. The weights are obtained from structural steelwork tables.

*Taking off*
The measurement should proceed in a logical sequence possibly following the inclusion of a general description of the work, which would include reference to location drawings. Fabrication will precede erection in the following format.

(1) *Fabrication of framing*
    (a) Columns or stanchions and their associated fittings.
    (b) Main beams and their associated fittings.
    (c) Subsidiary beams and their associated fittings.
    (d) Roof trusses and their associated fittings.

(2) *Erection*
  (a) Erection of framing (permanent erection on site).
  (b) Trial erection of framing, if required, when it would be desirable to state the location.
  (c) Any site works items relating to erection.

(3) *Surface preparation and treatment*

These items would be followed by the measurement of any associated concrete casings to columns, beams and the like, including formwork and any fabric or other wrapping.

DRG. 9                                    STRUCTURAL   STEELWORK
                                                      Sheet  1

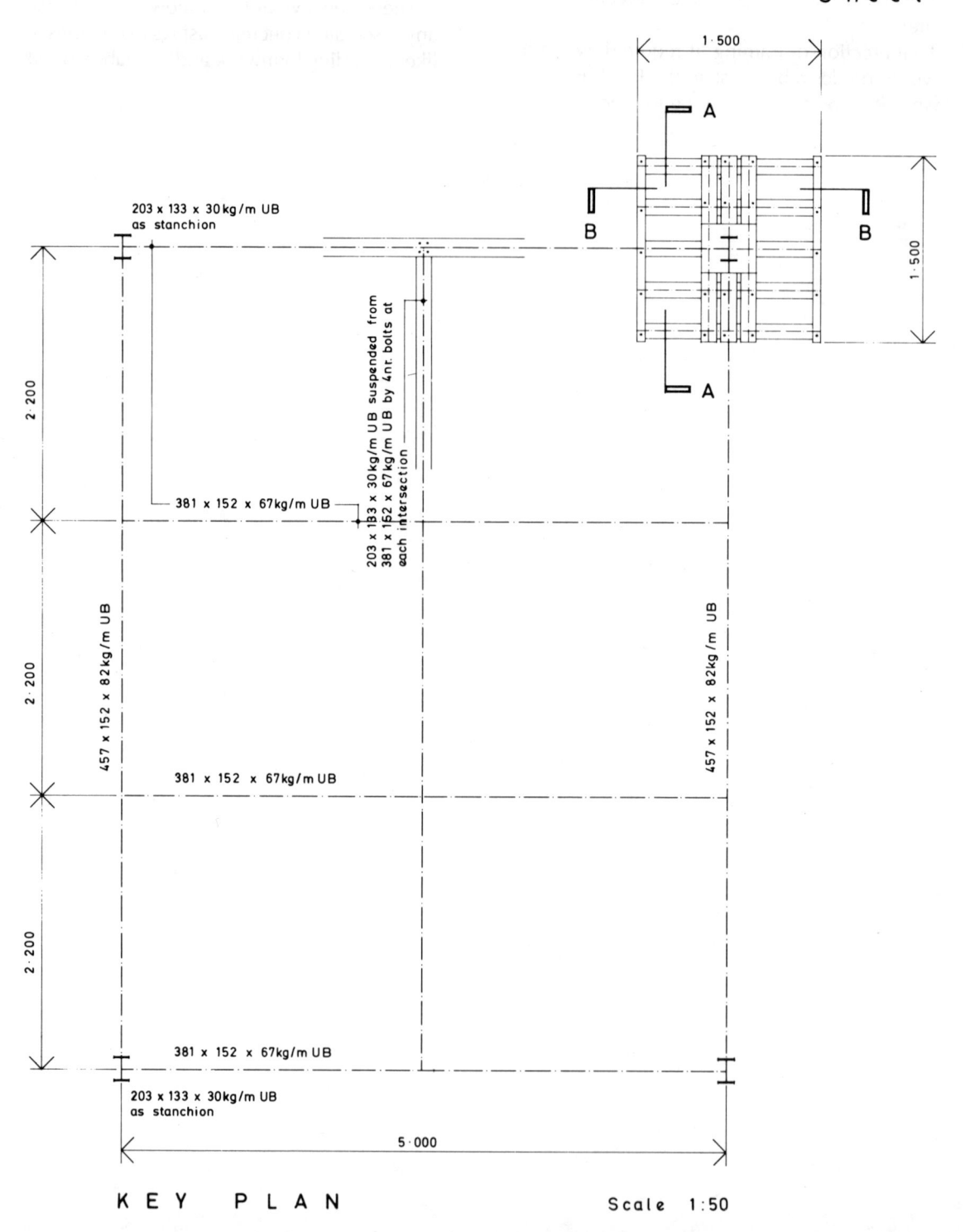

203 x 133 x 30 kg/m UB
as stanchion

203 x 133 x 30kg/m UB suspended from
381 x 152 x 67kg/m UB by 4nr. bolts at
each intersection

381 x 152 x 67kg/m UB

457 x 152 x 82kg/m UB

457 x 152 x 82kg/m UB

381 x 152 x 67kg/m UB

381 x 152 x 67kg/m UB

203 x 133 x 30kg/m UB
as stanchion

K E Y   P L A N                                    Scale  1:50

DRG. 9

STRUCTURAL STEELWORK
Sheet 2

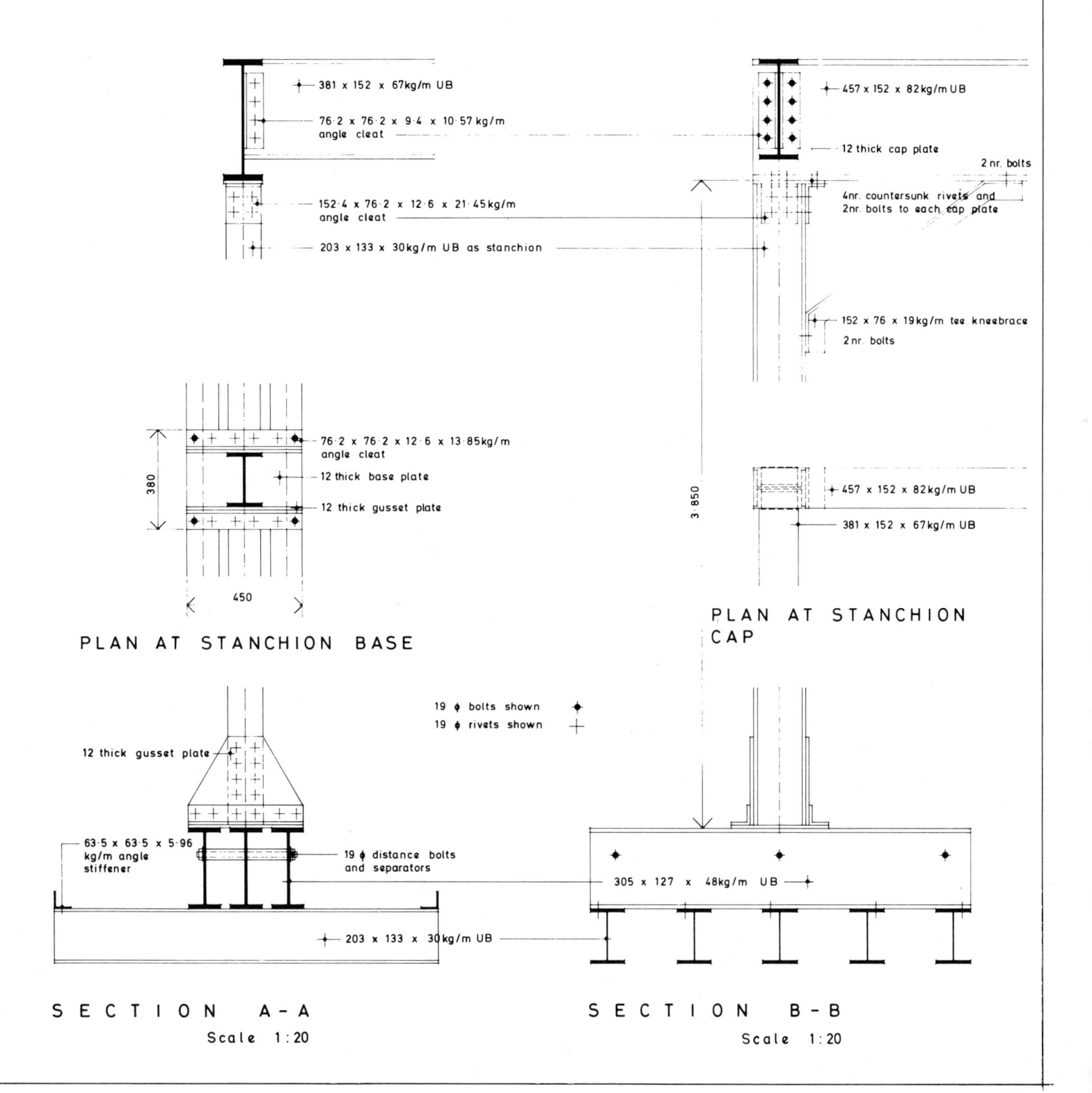

- 381 x 152 x 67kg/m UB
- 76·2 x 76·2 x 9·4 x 10·57 kg/m angle cleat
- 152·4 x 76·2 x 12·6 x 21·45kg/m angle cleat
- 203 x 133 x 30kg/m UB as stanchion

- 457 x 152 x 82kg/m UB
- 12 thick cap plate
- 2 nr. bolts
- 4nr. countersunk rivets and 2nr. bolts to each cap plate
- 152 x 76 x 19kg/m tee kneebrace
- 2 nr. bolts

- 76·2 x 76·2 x 12·6 x 13·85kg/m angle cleat
- 12 thick base plate
- 12 thick gusset plate

380
450
3·850

- 457 x 152 x 82kg/m UB
- 381 x 152 x 67kg/m UB

## PLAN AT STANCHION BASE

## PLAN AT STANCHION CAP

- 12 thick gusset plate
- 63·5 x 63·5 x 5·96 kg/m angle stiffener
- 19 ⌀ distance bolts and separators
- 203 x 133 x 30 kg/m UB

19 ⌀ bolts shown
19 ⌀ rivets shown

- 305 x 127 x 48kg/m UB

## SECTION A-A
Scale 1:20

## SECTION B-B
Scale 1:20

88

## STRUCTURAL STEELWORK
### STEEL FRAMED BUILDING

The wk. comprises a stl. frd. single storey bldg. 6.60 m x 5.00 m x 4.31 m hi. above g.l. as shown on Location Drawing 9 of riveted fabrication w. bolted site connectns. The steelwk. is to conform to BS 4.

```
              3.850
                457
              4.307
```

| | | |
|---|---|---|
| | | A general description of the work has been given to assist the estimator, although it is not mandatory under SMM 7. SMM G10.P1 requires the following information to be shown either on location drawings or on further drawings to accompany the bill of quantities. |
| | | (a) position of the work in relation to other parts of the work and of the proposed buildings; |
| | | (b) types and sizes of structural members and their positions in relation to each other; |
| | | (c) details of connections or of the reactions, moments and axial loads at connection points. |

### Grillages

| | | | |
|---|---|---|---|
| ✳ | ___ | kg | Structl. stl. framg., fabricatn., grillages. |

This item will subsequently be reduced to tonnes for transfer to the bill (SMM G10.1.5.0.0).

✳ comprisg. total wt. of follg. components

| 4/5/ | 1.50 | 203 x 133 x 30 kg/m U B. |
|---|---|---|
| 4/2/ | 1.50 | 63.5 x 63.5 x 6.2 x 5.96 kg/m L stiffener. |
| 4/3/ | 1.50 | 305 x 127 x 48 kg/m U B. |

The components making up the grillage are then measured and described, stating the weight per metre, for subsequent weighting under the omnibus grillages item, already entered on the dimensions paper. The black bolts connecting the universal beams in the grillages are deemed to be included and hence do not require separate measurement (SMM G10.C1).

```
          len. of septrs.
     o/a len.        320
     less thickness of
          centre web   9
                   2)311
     individl. lens. 156
```

However, the separators need weighting as associated fittings to the grillage.

Calculation of lengths of separators in waste.

9 . 1

---

EXAMPLE 9

| 4/3/2/ | 0.16 | Stl. separators, 19 int. ⌀, 25 ext. ⌀ x ___ kg. |
|---|---|---|

End of grillages

### Cols.

| | | |
|---|---|---|
| ⌽ | ___ kg | Struct. stl. frmg., fabricatn., cols. |

⌽ comprisg. total wt. of follg. membrs. & fittgs.

```
                3.850
    less plates 2/12   24
                3.826
```

| 4/ | 3.83 | 203 x 133 x 30 kg/m U B. |
|---|---|---|

### Bases

| 4/ | 0.45 | 12 th. base plate |
|---|---|---|
| | 0.38 | x ___ kg |

| 4/2/ | 0.45 | 76.2 x 76.2 x 12.6 x 13.85 kg/m L cleat. (base of col. |
|---|---|---|

```
   width at base  450
   width at top   133
                2)583
   av. width of topd. part 292
```

| 4/2/ | 0.45 | 12 th. gusset plate |
|---|---|---|
| | 0.08 | x ___ kg |
| 4/2/ | 0.29 | |
| | 0.26 | |

The calculations, although seemingly rather trivial, should be inserted in full in waste to show the precise method of computation.

The measurement of the stanchions follows the classification of columns in SMM G10.1.1.0.0. In calculating overall lengths, no deductions are made for splay cuts or mitred ends (SMM G10. M3). The associated fittings are weighted together with the members to which they are attached.

A different symbol has been used for each component of the structure to separate the fabrication items for the various members, coupled with the associated fittings, which are weighted together in tonnes. The same symbols can be used for collating the single weighted erection item under SMM G10. 2.2.0.0, although this may appear a rather protracted way of grouping the items. Its value would become more apparent if there were more components involved. There are various ways of entering the dimensions of steelwork and the method used in this example is not necessarily universal. The main objective is to clearly identify and group all the appropriate items as required and to bill them in tonnes.

9 . 2

## STRUCTURAL STEELWORK (Contd.)

| | | | |
|---|---|---|---|
| | | Cap plates | |
| | | col. | 203 |
| | | cleat | 76.2 |
| | | | 279.2 |
| 4/ | 0.28 | 12 th. cap plate. | |
| | 0.15 | x ___ kg | |
| 4/2/ | 0.14 | 152.4 × 76.2 × 12.6 (web | |
| | | × 21.45 kg/m L | |
| 4/ | 0.15 | cleat. (flanges | |

End of cols.

Roof Beams

| ⊠ | ___ kg | Struct. stl. frmg., fabricatn., beams. |

⊠ comprisg. total wt. of follg. membrs. & fittings.

len. of main beams
3/2.200  6.600
add ends 2/½/203  203
              6.803

| 2/ | 6.80 | 457 × 152 × 82 kg/m UB. (main beams |

| 4/ | 5.00 | 381 × 152 × 67 kg/m UB. (cross beams |

| 4/2/2/ | 0.30 | 76.2 × 76.2 × 9.4 × 10.57 kg/m L cleat. (cleats connecting beams |

Fittings include caps, bases, gussets, end plates, cleats, stiffeners, distance pieces and the like. They are all given by weight and added to the appropriate member, unless they are of a different type and/or grade of metal.
The width of the tapered part of the gusset plate is averaged in waste and the height is scaled from the drawing.
The last fittings to be measured are the angle cleats providing support to the cap plate.
It is good practice to indicate when the end of the items associated with each type of member is reached.
Fabrication of both main and subsidiary roof beams are collectively classified as beams in accordance with SMM G10.1.2.0.0.
Half the depth of each end column is added to the dimensions of the spans as figured on the drawing, to give the total length of the main roof beams.
Each member is suitably marked on the drawing following taking off, so that on completion all members will be marked. This gives the weighted quantity for billing purposes (length × weight per metre, converted to tonnes).
This method is more elaborate than that adopted by some quantity surveyors, but does clearly identify all the items involved.
The roof beam is followed by the associated fittings items as SMM G10.M1, and these are weighted together in tonnes. The length of the cleat has been scaled.

9.3

---

len. of centre beam
3/2.200  6.600
add ends 2/½/152  152
                  6.752

| | 6.75 | 203 × 133 × 30 kg/m UB. (centre beam |

len. of knee braces
                860
add ends 2/150  300
              1.160

| 4/ | 1.16 | 152 × 76 × 19 kg/m tee (knee braces |

End of beams

Erectn.

| ✱ | ___ kg | Framg. perm. erectn. on site. (grillages |
| ⊕ | ___ kg | (total wt. ___ t) (cols. |
| ⊠ | ___ kg | (beams |

Painting
(taken above tops of gusset plates to column bases)

Girths

| Cols. & | | Main |
| Centre Beam | | Beams |
| 2/203 | 406 | 2/457 | 914 |
| 4/133 | 532 | 4/152 | 608 |
| | 938 | | 1.522 |

Half the flange width of the main beams is added to give the total length of the centre beam.

This item will provide the weight of the centre roof beam for inclusion in the bill.

All black bolts, nuts and washers required for the beam connections are deemed to be included in the weighted fabricated members and associated fittings (SMM G10.C1), and this includes the bolts used for fixing the suspended centre beam to the cross beams.

The dimensions are scaled from the drawing in the absence of figured dimensions.
The kneebraces are taken as fittings to be added to the beams, but alternatively they could be added to the weight of the columns.

The total weight of all the structural steel framing is taken for permanent erection on site and billed in tonnes as SMM G10.2.2.0.0.
It is good practice to show the location of the members in waste.
The painting is taken above the gusset plates at the bottoms of the columns (stanchions) assuming that these will be encased in concrete in the foundations.
The girths of the columns and beams consist of twice the depth and four times the breadth, as there are two sides to be painted to each flange.

9.4

## STRUCTURAL STEELWORK (Contd.)

| | | | | |
|---|---|---|---|---|
| | | Cross Beams | Knee braces | The knee braces merely require twice the depth and breadth respectively. |
| | | 2/381  762 | 152 | |
| | | 4/152  608 | 76 | |
| | | 1·370 | 2/ 228 | |
| | | | 456 | |

| | Col. Len. |
|---|---|
| | 3·850 |
| less gussets | 350 |
| | 3·500 |

| | | | | |
|---|---|---|---|---|
| 4/ | 3·50 | Paintg., structl. | (cols. | Site painting to structural metalwork is measured the exposed girth in $m^2$ as the girth is $> 300\,mm$, in accordance with SMM M60.5.1.1.0, and has been classified as external work, assuming that it is to be painted prior to being enclosed (SMM M60.D1). |
| | 0·94 | metalwk., g.s., ext., girth > 300, prime & ③. | | |
| 4/ | 1·16 | | (knee-braces | |
| | 0·46 | | | |
| 2/ | 6·80 | | (main beams | |
| | 1·52 | | | |
| 4/ | 5·00 | | (cross beams | Painting to structural metal-work 5·00 m or more above floor level is measured separately in 3·00 m stages (SMM M60.5.1.1.4–5), but none of the steelwork in this example comes into that category. The height of structural metalwork is measured to the highest point of the members in the stated height range (SMM M60.M8). |
| | 1·37 | | | |
| | 6·75 | | (centre beam | |
| | 0·94 | | | |

### TRUSSES
(framg., fabricatn.)

12 nr trusses in m.s. ea. 12·00 m lg. & 2·50 m hi. w. riveted jts. (ref. 1A, Drawing S4) & ea. comprisg :

| | | | | |
|---|---|---|---|---|
| (a) | 4 nr | 76·2 x 63·5 x 6·2 x 6·56 kg/m Ls, ea. 6·50 m lg.; total wt. 0·17 t. | Total wt. of membrs. 3·96 t. | It is considered helpful to the student to complete the worked examples on structural steelwork with 12 roof trusses using a different method of tabulation to that used for structural framework. The heading gives details of the number, overall dimensions and construction of the trusses and refers to the detailed drawing. The component members are then listed giving the dimensions, weight and type of member. |
| (b) | 4 nr | Ditto., 3·20 m lg.; total wt. 0·08 t. | | |
| (c) | 4 nr | 63·5 x 50·8 x 6·2 x 5·35 kg/m Ls, ea. 2·50 m lg.; total wt. 0·05 t. | | |
| (d) | 2 nr | Ditto., 3·00 m lg.; total wt. 0·03 t. | | |

The total weight of steel in the members in all 12 trusses is entered in the quantity column in the bill, and is computed as follows :

9.5

---

The members comprising the trusses will be followed by a weighted item of fittings in tonnes, embracing the angle cleats and gusset plates forming the connections, which are then added to the weight of the truss members, to give the total weight of the trusses.

$((a) + (b) + (c) + (d)) \times 12$
$= (0.17 + 0.08 + 0.05 + 0.03) \times 12$
$= 0.33 \times 12 = \underline{3.96\,t}$

The trusses (framing, fabrication) item, classified as SMM G10.1.8.1.0, is followed by the associated fittings which are weighted up with the truss members (SMM G10.M1).
Purlins are measured as a separate item (SMM G10.1.4.0.0).

9.6

# 6 WOODWORK AND METALWORK

## TIMBER PARTICULARS

The measurement of woodwork falls into a number of different categories, such as carpentry/timber framing/first fixing (G20), unframed isolated trims/skirtings/sundry items (P20), timber windows/rooflights/screens/louvres (L10), timber doors/shutters/hatches (L20) and timber stairs/walkways/balustrades (L30). It is necessary to distinguish between sawn and wrought timber and the kind and quality of timber should be specifically stated, in order that the estimator can determine a realistic price for the supply of the timber and estimate with a reasonable degree of accuracy the relative ease or otherwise of working the timber (SMM G20/L10/L20/L30.S1). For example, if the bill description merely refers to 'oak' then the contractor will be justified in providing the cheapest variety available.

The reader is also referred to the other provisions of SMM7 prescribing the timber particulars to be given in the bill. For instance, the preservative treatment applied as part of the production process, such as the full cell process; surface treatment applied as part of the production process, such as shop priming; the selection and protection for subsequent treatment, such as 'left in the white' or to receive transparent finishes (SMM7 Code of Procedure G20.S6); and matching grain or colour.

All timber sizes are deemed to be nominal sizes unless stated as finished sizes (SMM G20.D1). There can be a variation in planing margins and hence nominal sizes are generally preferable.

It is necessary to state the method of jointing or form of construction where not at the discretion of the contractor (SMMG 20.S9), as he could, for instance, use butt joints instead of lapped joints on wall plates.

Long timber members that are required to be in one continuous length and exceed the normal maximum imported lengths are kept separate. Thus those > 6.00 m in length are so described stating the length.

With carpentry items the work is deemed to include labours on the timbers (SMM G20.C1). With linear joinery items such as skirtings, the dimensioned overall cross-section description will pick up the labours, such as chamfers and mouldings. Stopped labours are included in the item description and enumerated. Ends, angles, mitres, intersections and the like are deemed to be included in the work without the need for specific mention, except on hardwood items > 0.003 m$^2$ sectional area, when they are separately enumerated (SMM P20.C1). To illustrate the relative significance of this limit, a 25 × 125 mm member has a sectional area of 0.0031 m$^2$.

Members of identical cross-section but varying shapes constitute separate items. For example, 100 × 75 mm softwood frames could be in a variety of different shapes resulting from variations in rebates and other labours. These will influence machine settings and hence costs and need to be separated in the bill. Items which do not have a constant cross-section are so described, stating the extreme dimensions (SMM P20.M1).

## CARPENTRY/TIMBER FRAMING

A number of examples of the measurement of carpentry timbers in floors, stud partitions and roofs are contained in *Building Quantities Explained*, with explanatory notes describing the main implications of the relevant clauses of SMM7 and their application. All these timbers are measured in metres with a dimensioned description, under the classifications listed in SMM G20.1–10. The lengths measured are the extreme lengths with no additions for running joints.

The different pitched roof timbers, such as rafters, hip and valley rafters, ceiling joists, ridge boards, struts, purlins, binders and bracings are no longer separated and described, but are grouped together as pitched roof members with a dimensioned description (SMM G20.D6). A typical example would be 'roof members, pitched, 50 × 100 sawn softwood'. Another alternative approach might be '50 × 100 sawn softwood in pitched roof members'.

Flat roof members include joists and bearers (SMM G20.D5). Plates include bearers and partition members include struts and noggings (SMM G20.D3 and 4). It will be noted that joist strutting is classified as herringbone or block and is measured over the joists (SMM G20.10.1–2.0.0 and G20.M1), whereas noggings in stud partitions are measured the lengths between the studs. The kind and quality of timber is often given in a preamble clause, and a typical entry could be 'impregnated sawn softwood grade S2/MGS'.

## FLOORING AND ROOF BOARDING

Boarding to floors and roofs is measured in $m^2$, when > 300 mm wide, giving a dimensioned description, including the method of jointing and fixing where not at the discretion of the contractor, and the appropriate classification as SMM H21/K20.1–5.1.1.0. These items include selection and protection for subsequent treatments (SMM K20.S8), but the final treatment, such as polishing, will be dealt with under section M60 (Painting/clear finishing). No deduction is made for voids $\leq 0.50$ $m^2$, but this allowance does not apply to openings or wants at the boundaries of measured areas (SMM General Rules 3.4). Narrow widths ($\leq 300$ mm) are given in metres, while small areas ($\leq 1.00$ $m^2$) are enumerated. Floor boarding in door openings may be regarded as an extension of the boarding in the adjoining room, but the bearers will require separate measurement, classified as floor members. Work is deemed to include labours (SMM K20.Cla). Worked examples of the measurement of boarded floors are given in *Building Quantities Explained* (examples VII and IX).

Supports to flat roofs, which include firrings, drips, rolls and upstands, are given in metres with a dimensioned overall cross-section description (SMM G20.13.0.1.1 and 4). A typical description for a roll to a lead flat roof would be 'individual support, 30 diameter × 50 high, irregular shaped area with semi-circular top, wrought softwood'. Worked examples of the measurement of flat roofs are given in *Building Quantities Explained* (examples XIV and XV) and in example 15 in this book.

## LININGS AND CASINGS

Sheet linings and casings are measured in a similar manner to floor boarding. Particulars shall be given of the type of sheet (plywood, chipboard, fibreboard, asbestos cement, plastics and applied finishes), quality, thickness, method of jointing, method of fixing, nature of background and other relevant details as SMM K13.S1–13. The work is deemed to be internal unless described as external (SMM K13.D1). Note the separate treatment of work to isolated beams, columns and abutments (SMM K13.6–8), and that work to ceilings and beams over 3.50 m above floor, except in staircase areas, is so described, stating the height in further 1.50 m stages (SMM K13.M2). Labours to sheet linings and casings are deemed to be included (SMM K13.C1a). Access panels are enumerated as extra over the work in which they occur, with a dimensioned description (SMM K13.12.1.1.0).

## COMPOSITE ITEMS

*Fitments*
Composite items comprise kitchen fitments and the like which are fabricated off site and are normally enumerated with a component drawing reference or a dimensioned diagram (SMM N10.1.1–2.0.0), although it is permissible to use other appropriate rules from SMM7, provided the rules so used are identified (SMM N10.M1). Installation on the site is normally deemed to be included, as is the breaking down of composite items for transport and subsequent reassembly (SMM General Rules 9.1). Ancillaries not provided with the fittings are separately enumerated, stating the type, size and method of jointing (SMM N10.5.1.1.0). Hence fitments fabricated off site come within the category of composite items, whereas fittings like a length of shelving would not. The latter would be a separate linear item with a dimensioned overall cross-section description as SMM P20.3.1.0.0. The term 'composite' is also a little misleading as, in normal constructional phraseology, this implies an item made up of more than one material, such as a combination of wood and steel. It is purely coincidental that the fixed benching in example 11 is made up of both metal and timber sections, and even there the two parts have been separated.

Cover fillets and other associated work which do not form an integral part of the composite item, and are supplied separately and fixed after assembly, are given separately as linear items with a dimensioned overall cross-section description as SMM P20.2.1.0.0. Labours are normally included in the descriptions of the items on which they occur, and ends, angles and intersections are deemed to be included in the Items without the need for their specific mention, except on hardwood items > 0.003 $m^2$ sectional area (SMM P20.C1).

Example 11 illustrates the method of approach to two composite items, one of metalwork and the other of timber, together making up a bench unit. The enumerated items are accompanied by relatively brief descriptions giving the overall dimensions and referring the estimator to the component drawing as described in SMM N10.1.1.0.0, for full constructional particulars.

To assist the student by giving more practice in the measurement of component members in both metal and timber, the detailed measurement is included in example 11. In each case a logical sequence is adopted with the measurements built up in waste from figured dimensions on the drawing as far as possible, and following the requirements of SMM7 wherever practicable. In some instances with the metal base unit, it is difficult to find clauses in the Standard Method that are entirely relevant, and some improvisation is necessary. It will, however, be appreciated that SMM7 cannot possibly deal with every situation that can arise in practice and that its main purpose is to establish general guidelines. For example, the pedestal fittings are taken as enumerated omnibus fully described items, while the base channel and angle to the benchtop are measured as linear items with the method of jointing or form of construction included in the item description.

In the measurement of the timber benchtop, the edgings, nosings, fascia and upstand are measured in metres, giving the material, and a dimensioned overall cross-section description. Only in the case of hardwood members > 0.003 $m^2$ in sectional area will it be necessary to enumerate ends, angles, mitres and intersections (SMM P20.C1). The chipboarding is measured in accordance with the rules for sheet linings, giving the particulars listed in SMM K13. Where the width > 300 mm it is measured in $m^2$, otherwise in metres. Labours are deemed to be included as SMM K13.C1a. Note the approach adopted for the measurement of the modesty panels, the ends to the benching, associated metal items and the leather facing. Full use has been made of sub-headings, preliminary calculations and locational notes to make the dimensions as comprehensive and foolproof as possible.

*Doors and adjoining screens*

Example 10 shows the method of measuring glazed doors and adjoining lights and glazed screens. The sequence of measurement of the doors and associated lights follows the order of construction on the site, starting with the door frame followed by the door, as distinct from the more commonly adopted approach of taking the door first, on the basis that the door size determines the frame/lining and opening sizes. The door frame provides a good subject for a dimensioned diagram as described in SMM General Rules 5.3, showing the component members with their functions and sizes and the overall dimensions. The component members are then measured in metres, giving dimensioned overall cross-section descriptions and the number of sills and transoms (SMM L20.7.1–5.1.0).

The method of fixing wood frames, where not at the discretion of the contractor, is included in the component description (SMM L20.S8), but mortices in masonry are deemed to be included in the masonry items (SMM F21.C1c).

Doors are enumerated with a dimensioned diagram as SMM L20.1.0.1.0, and each leaf is counted as one door (SMM L20.M2), and fitting and hanging are deemed to be included (SMM L20.C1). In this particular case the dimensioned diagram can be extracted from Drawing 10. Enumerated items for fixing ironmongery, which include the nature of the background to which they are fixed, are inserted. The supply of ironmongery is often covered by a prime cost sum accompanied by a suitable description and with provision for the addition of profit as SMM A52.1.1–2.1.0.

The measurement of the glazing to the doors entails calculating the pane sizes in waste to obtain the dimensions for insertion in the dimension column and to determine the pane size classification as SMM L40.1.1.1–2.0. Full particulars of the glass and method of glazing are required as prescribed by SMM L40.S1–5. The student should note particularly that glazing beads are included in the door item, but that their fixing is incorporated in with the glass. Panes of irregular shape are so classified and measured according to the smallest rectangular area from which the pane can be obtained (SMM L40.M3). Where panes do not exceed 0.15 $m^2$ in area, the number of panes is stated, and if there are fifty or more identical panes they shall be given separately stating their number and size, to permit the savings in cost resulting from the extensive repetition of similar items to be obtained (SMM L40.1.1.1–2.1).

When measuring the painting, it is best to follow the same sequence as for the joinery, as there will then be less risk of omissions occurring. Painting to door frames, linings and associated mouldings, such as architraves, and sills is measured separately from the glazed doors and accompanying lights, as general surfaces in accordance with SMM M60.1.0.2.0. Where the girth is ≤ 300 mm. it is measured in metres (SMM M60.1.0.2.0). All work is deemed to be internal unless otherwise described (SMM M60.D1). The description of the materials, base, preparatory work and painting shall include the particulars listed in SMM M60.S1–8.

The painting of the glazed doors and painting of glazed screens are so classified in accordance with SMM M60.2 & 4.1–4.1.3, giving the appropriate pane size classification. Where panes of more than one size occur the sizes are averaged (SMM M60.M6). It is customary to times the area of glazed doors by $1\frac{1}{9}$ to cover the opening edges and any mouldings. Painting of glazed work is deemed to include additional painting to the surrounding frame caused by opening lights, cutting in next glass and work on glazing beads, butts and fastenings attached thereto (SMM M60.C4). Work is described as multi-coloured when more than one colour is applied to an individual surface, except on walls and piers and on ceilings and beams (SMM M60.D2), and the application of this provision is illustrated in the SMM7 Code of Procedure.

*Windows*

The measurement of windows of various types is described and illustrated with worked examples in *Building Quantities Explained*. They are described and enumerated including frames and window surrounds, and generally accompanied by dimensioned diagrams.

*Glazed screens*

Example 10 illustrates the approach to the measurement of glazed screens, commencing with a dimensioned diagram as SMM General Rules 5.3. The screens would probably be fabricated off site and delivered with glazing beads temporarily bradded into position. The screens are enumerated and described in accordance with SMM L10.5.0.1.0, stating the overall dimensions, materials used, including glazing beads, and method of fixing where not at the discretion of the contractor. All timber sizes are deemed to be nominal sizes, unless stated as finished sizes (SMM L10.D1), as in the case of the glazing beads. The description is substantially reduced by referring to Drawing 10 for full details of the screens. This is followed by the bedding and pointing of frames.

The measurement of glazing follows a similar approach to that adopted for the glazed doors. This is followed by the painting where it is necessary to measure the varnishing to both sides of the screens, including uprights and heads, in $m^2$ as SMM M60.2.1–4.1.0, and to the sills in metres as general surfaces in accordance with SMM M60.M7. Painting/varnishing to heavily moulded surfaces would be classified as irregular surfaces.

*Staircases*

Staircases are enumerated as composite items with a dimensioned description or component drawing (SMM L30.1.1–2.0.0). It will be easier for the estimator to price from the component drawing than a dimensioned description. In the absence of full component details, a provisional sum would be entered in accordance with SMM A54.1.1.0.0. A worked example of the measurement of a timber staircase is provided in example XXII in *Building Quantities Explained*.

# FOR READER'S NOTES

# DRG. 10    DOORS AND ADJOINING LIGHTS COMPONENT DETAILS

## FRONT ELEVATION

## SECTION X-X

- copper roofing with welted seams and underlay
- 2·400
- 2·150
- 1·800
- stone step
- 600 x 150 concrete foundation
- 50 screed on 50 insulating concrete on 1000 grade visqueen on 86 concrete on 150 hardcore (min.)

## PART PLAN   Scale 1:100

- 660 / 660
- 3·370
- mat well
- 1·760
- 2·550
- line of edge beam

**Notes:** All joinery to be in selected teak, prepared and finished with three coats of polyurethane clear finish

## PART PLAN OF GLAZED FRONT

- dpc
- ex. 150 x 75 side frame
- 31 x 13 (fin.) glazing bead with brass cups and screws
- mastic pointing
- Bulwell stone face
- ex. 150 x 62 mullion
- ex. 150 x 62 side frame bolted to corner post with 4 nr. 10 ⌀ coach bolts
- ex. 150 x 120 corner post
- ex. 200 x 50 rail
- 25 dowel
- ex. 150 x 75 jamb
- ex. 150 x 50 door stile

## SECTION B-B

- ex. 75 x 150 filler piece scribed to curve of ceiling
- line of ceiling (curved in one direction)
- ex. 150 x 75 head
- 6 polished plate glass bedded in putty
- edge beam to roof
- ex. 150 x 75 head
- 31 x 13 bead
- 6 polished plate glass bedded in putty
- ex. 200 x 75 sill
- ex. 150 x 25 sill board
- dpc
- floor level

## SECTION A-A   Scale 1:10

- ex. 150 x 75 head
- 31 x 13 (fin.) glazing bead with brass cups and screws
- ex. 150 x 75 transom
- ex. 150 x 50 top rail
- 31 x 13 (fin.) glazing bead with brass cups and screws
- 6 polished plate glass bedded in plastic glazing strip
- ex. 200 x 50 lock rail
- ex. 225 x 50 bottom rail
- patent weather bar
- 30 x 25 brass angle mat well
- 300 x 150 Bulwell stone step

## DOORS AND ADJOINING LIGHTS

Inr dr. frame set as dimnsd. diagrm. nr. 1

```
         ┌─ 150 x 75 head
   ┌─────┴─────┐
   │  ┌─────┐  │
   │  │150x75 transom
2.400 │  ├─────┤
   │  │150 x 62 side posts
   │  │←150 x 75 jambs→
   │  │200 x 75 sill
   └──┴─────┴──┘
      ← 2·550 →
```

Dimnsd. diagram nr 1

| | | |
|---|---|---|
| 2/ | 2·40 | Jambs 150 x 62 wrot teak spld. & reb.  (side posts |
| 2/ | 2·15 | Jambs 150 x 75 wrot teak 2ce spld. & 2ce reb., fxd. w. m.s. dowels, 25 ø x 150 lg. |
| | 2·55 | Head 150 x 75 wrot teak spld. & reb. &  Transom 150 x 75 wrot teak 2ce spld. & 2ce reb.  (In 1 nr) |

```
  o/a. len. of frt.   2·550
  less drs.           1·760
                    2) 790
                       395
  add 2 faces to both
  corner posts 2/2/100 400
                       795
```

| | | |
|---|---|---|
| 2/ | 0·80 | Sill 200 x 75 wrot teak wethd., reb., grvd. & thro.  (In 2 nr) |

End of dr. fr. set

| | | |
|---|---|---|
| 2/ | 0·80 | Bed wd. fr. in c.m. (1:3) & pt. in mastic o.s. |

The measurement of this work has been subdivided into the doors and adjoining lights and frames; and the lights to the screens on the return facades, as a different form of measurement applies in each case. With the doors and associated work, it has been decided to follow the order of construction on site and to commence with the measurement of the door frame set, even although most quantity surveyors start by measuring the door(s) and that this practice is adopted in 'Building Quantities Explained'. The door frame set is grouped together under a suitable heading as SMM L20.7.1-5.1.0 and this is an instance when a dimensioned diagram will prove very helpful (See SMM General Rules 5.3).
Some surveyors may regard the framing as a composite set (SMM L20.7.6).
Jambs, heads, sills, mullions, transoms and the like are each measured in metres, with a dimensioned overall cross-section description. The number of lengths of sills, mullions and transoms is also to be stated.
Particulars of the timber are to be given as SMM L20.S1-10 and all sizes shall be deemed to be nominal unless stated as finished sizes (SMM L20.D1).

The method of fixing is given where not at the discretion of the contractor (SMM L20.S8).

10.1

## EXAMPLE 10

| | | |
|---|---|---|
| | 2·55 | Filler piece, 75 x 150 wrot teak scribed to curve of clg. |

Corner Posts

| | | |
|---|---|---|
| 2/ | 2·40 | Corner posts 150 x 120 wrot teak 4 times spld., ea. fxd. w. g.m.s. coach bolts, 10 ø x 80 lg., inc. counter-skg. & pelletg.  (In 2 nr) |

Doors
```
              width
         2) 1·760
             880
  add ½ reb. on stile  10
             890
             ht.
             2·150
  less fr.    55
             2·095
```

| | | |
|---|---|---|
| 2/ | 1 | Dr. as component drg. 10, 890 x 2·095 x 50 th. wrot teak 2 pan. chfld. o.s.w. both pans. open & reb. for glass (m/s), inc. 31 x 13 (fin.) teak glazg. beads. |

Provide the P.C. Sum of £     for supply of patent weather bar, delvd. to site.
&
Add for profit.

10.2

Bedding and pointing frames is measured in metres (SMM L20.8-10). Sills are taken around the corner posts to join the screens. There is no provision in SMM L20 for the measurement of end labours to framed members, such as splayed ends and mitres. This is a special item which needs to be adequately described.
It is considered convenient to insert the two corner posts here, including the fixing by four bolts to each side of each post, as they do not form an integral part of either the door frame set or the adjoining screens.

The dimensions of each door are calculated in waste from the overall figured measurements on the drawings, making the relevant adjustments.
Doors are described and enumerated (each leaf being counted as one door), and reference made to a dimensioned diagram for the necessary particulars (SMM L20.1.0.1.0). The door description should include such essential details as overall dimensions, number of panels, number of glazed panels and glazing beads. Fitting and hanging are deemed to be included with the doors (SMM L20.C1).
Another design of door is measured in Example XXI of 'Building Quantities Explained'.

The supply only of fittings can be covered by a prime cost sum, with provision for the addition of profit as SMM A52.1.1-2.1.0.

## DOORS AND ADJOINING LIGHTS (Cont'd)

| | | | |
|---|---|---|---|
| | 1 | Fix patent stainless stl. weather bar, 1·76 m lg., w. stainless stl. scrs. to hwd. plugs in mors. in Bulwell stone. | The fixing of the patent weather bar will form a separate enumerated item, including the nature of the background to which it is to be fixed and the method of fixing. |

```
                    Glazg.
                    width
less 2 stiles less   890
rebs. 2/150-13       274
                     616
                     ht.
less                 2·095
top rl. (less reb.)
150-13       137
lock rl. (less rebs.)
200-26       174
bott. rl. (less reb.)
225-13       212    523
        total ht.   1·572
        less bott. pane  690
        ht. of top pane  882
```

The elevational drawing is drawn to too small a scale from which to accurately measure glass sizes and they are not figured on the larger scale plan and section. Hence it is necessary to calculate the glass dimensions by taking the overall door dimensions and deducting the width of members, less the depth of rebates (13 mm). The height of the bottom pane is then scaled to arrive at the height of the top pane by deduction from the total height of glass.

| | | | |
|---|---|---|---|
| 2/ | 0·62 | | |
| | 0·69 | Glazg., stand. plain glass, panes area 0·15 - 4·00 m² pol. plate glass, 6 th., to BS 952 to hwd. w. hwd. glazg. beads (m/s), secured w. brass cups & scrs. & bedded in neoprene strip. (bott. panes) | Full particulars of the glass and glazing, including pane area classification, are given as SMM L40.1.1.2.0 and L40.S1-5. |
| 2/ | 0·62 | | The glazing beads were included in the description of the door, but their fixing is covered here as prescribed by SMM L40.S4. |
| | 0·88 | (top panes) | If it was laminated glass it would be classified as special glass to a dimensioned description (size given) as SMM L40.3.1.1.0. With sealed glazed units, full specification particulars or the proprietary reference should be supplied. |

### Ironmongery

| | | | |
|---|---|---|---|
| 2/ | 1½ | Prs. of 100 brass butts & fxg. to hwd. | Units or sets of ironmongery are enumerated separately as fixing items, with the exception of butts which are supply and fix items. |

10.3

| | | | |
|---|---|---|---|
| | 1 | Fix only mors. lock & lever furn. in alum. alloy to hwd. & Fix only 150 lg. brass barrel bolt & soc. to hwd. | Fixing items are inserted for ironmongery items to the doors, including the nature of the background to which they are to be fixed (SMM P21.1.1.0.0). Ironmongery is deemed to include fixing with screws to match and preparing base to receive same (SMM P21.C1), and the mortice in the stone step is deemed to be included in the masonry item under SMM F21.C1c. |

There will be a comprehensive prime cost item for the supply of all ironmongery to items of joinery.

```
                Sidelights
                width
len. of sill to
  ea. light          395
less dr. jb.   40
side frame     50    90
        net. width  305
                    ht.
        dr. ht.    2·095
less sill (less reb.)  60
        net. ht.   2·035
```

The sidelights and toplight surrounding the doors will each be enumerated and described with a reference to the component drawing in a similar manner to the doors.

As the framing has already been measured, the overall dimensions will be taken between the rebates on the framing members. The overall length of the front is taken to the outside edge of the side frames and hence a deduction is required for each side frame. The sidelights have been described giving the overall dimensions (inside the enclosing frames) and include the glazing beads and middle rail. As with the doors, the glass is measured separately. The nearest SMM7 classification is screens (SMM L10.5.0.1.0), but this does not seem appropriate in this case as part of the framing is common with the doors and screens are measured in the same way as windows, including the frames, which makes for complications.

| | | | |
|---|---|---|---|
| 2/ | 1 | Sidelt. to drs. as component drg. 10, 305 x 2035 inside teak reb. frs. (m/s), div. into 2 panes for glass (m/s), & includg. 31 x 13 (fin.) teak glazg. beads & middle rail. | |

10.4

## DOORS AND ADJOINING LIGHTS (Contd.)

| | | |
|---|---|---|
| | Glazg. | |
| | | ht. |
| | | 2.035 |
| | less middle rail | |
| | (less rebs.) | |
| | 200 – 2/13 | 174 |
| | total ht. of glass | 1.861 |
| | ht. of bott. pane | |
| | to dr. | 690 |
| | add | |
| | dr. bott. rl. 212 | |
| | less sill  60 | 152 |
| | | 842 |
| | total ht. of glass | 1.861 |
| | less bott. pane | 842 |
| | ht. of top pane | 1.019 |

| 2/ | 0.31 | |
| | 0.84 | Glazg., stand. plain glass, panes area: 0.15 – 4.00 m², pol. plate, 6 th., to BS 952 to hwd. w. hwd. glazg. beads (m/s), secured w. brass cups & scrs. & beddg. in l.o. putty. (bott. panes) |
| 2/ | 0.31 | |
| | 1.02 | (top panes) |

| | Top light | |
| | | len. |
| | | 2.550 |
| | less side frames 2/50 | 100 |
| | net. len. | 2.450 |
| | | ht. |
| | o/a ht. | 2.400 |
| | less dr. & trans. 2.150 | |
| | hd.  50 | 2.200 |
| | net. ht. | 200 |

| 1 | | Toplight to drs. as component drg. 10, 2450 × 200, inside teak reb. frs. (m/s), & includg. 31 × 13 (fin.) teak glazg. beads. |

Glazg.

| | 2.45 | Glazg., stand. plain glass, panes area: 0.15 – 4.00 m², pol. plate, 6 th., to BS 952 to hwd. w. hwd. glazg. beads (m/s), secured w. brass cups & scrs. & beddg. in l.o. putty. |
| | 0.20 | |

*The dimensions of the glass in the panes to the sidelights are calculated in a similar manner to that adopted for the doors. It is necessary to measure the individual panes separately as they might be in a separate panes area classification, although as it happens they both fall into the same classification (0.15 – 4.00 m²).*

*This description is similar to that for the doors, except that the glass is bedded in putty in place of the neoprene glazing strips, as it is to be in fixed lights. Note the locational particulars in waste and keep a close watch on timesing figures to ensure their accuracy. The net dimensions of the toplight are calculated in waste, starting from the figured dimensions. In the case of deductions for framing members, adjustment has to be made for rebates to receive the glass. This forms another enumerated item in the same way as the sidelights. It is good practice to make reference to the component drawing, which will help the estimator in pricing. An alternative could be to incorporate the doors, sidelights and toplights in a single omnibus enumerated item, but on balance the approach adopted would seem to be the best. The glazing item follows the same approach as for the sidelights. The panes area classification falls into the same classification (0.15 – 4.00 m²), as the area is 0.49 m². Pocket calculators are a godsend in making this sort of computation.*

10.5

---

Paintg.

| 2/ | 2.40 | Clear finishg., gen. isoltd. surfs., hwd., gth. ≤ 300, 3 cts. of polyurethane clear varnish. (side posts) |
| 2/ | 2.15 | (jbs. |
| | 2.55 | (hd. |
| | 2.55 | & (trans. |
| 2/ | 0.80 | Ditto., ext. (sills |
| | 2.55 | (filler piece |
| 2/ | 2.40 | Ditto., ext. (corner posts |

| 2/1½/ | 0.89 | Clear finishg., glazd. drs., hwd., panes area: (drs. 0.10 – 0.50 m², gth. > 300, 3 cts. of polyurethane clear varnish. |
| | 2.10 | |
| | | & |
| | | Ditto., ext. |

| 2/ | 0.31 | Ditto., glazed screens, hwd., panes area: 0.10 – 0.50 m², gth. > 300, 3 cts. of polyurethane clear varnish (varnishg. to frs. m/s). (side lights |
| | 2.04 | |
| | 2.45 | (top light |
| | 0.20 | |
| | | & |
| | | Ditto., ext. |

| | | len. |
| | 5/660 | 3.300 |
| | add | |
| | 2 half side frs. ½/62 | 31 |
| | ½/75 | 38 |
| | total len. | 3.369 |
| | | ht. |
| | | 2.400 |
| | | 1.800 |
| | 2) | 4.200 |
| | av. ht. | 2.100 |

*Follow the same sequence for painting as for joinery. Painting and clear finishing to wood frames is measured in metres as general isolated surfaces as ≤ 300 mm girth as SMM M60.1.0.2.0. Work is deemed to be internal unless described as external (SMM M60.D1). The work is deemed to include rubbing down with glass, emery or sand paper and so this does not require inclusion in the description (SMM M60.C1). The kind and quality of materials, nature of the base, preparation and treatment are to be described as SMM M60. S1-6. Where panes of more than one size occur, then the sizes are averaged (SMM M60.M6). The painting of the glazed doors is so described, stating the appropriate panes area classification as SMM M60.4.2.1.0.*

*The area of the doors has been timesed by 1⅓ to cover the opening edges, which are deemed to be included in the case of opening lights (SMM M60.C4a). For varnishing purposes, the adjoining lights to the doors are described as screens, to cover the varnishing work against the glass. However, this would normally include varnishing the framing, and so it is important to state in the description that this work has already been measured to avoid duplication. The length of the screens is calculated in waste from the component dimensions and then checked against the overall figured dimensions on the drawing (3.370).*

10.6

| DOORS AND ADJOINING LIGHTS (Contd.) | | | | |
|---|---|---|---|---|

2nr screens

Dimensioned diagram showing screen: 2.400 high, 3.370 wide, 1.800 at narrow end; 150 × 75 head; 150 × 62 mullions; 150 × 62 side frame; 150 × 75 side frame; 200 × 75 sill.

Dimnsd. diagram nr 2

| | | | |
|---|---|---|---|
| 2/ | 1 | Teak glazed screen 3·37 m lg. w. av. ht. of 2·10 m in accordance w. dimsd. diagrm. nr 2 & the component details shown on Drg. 10, & includg. all frmg. & 31 × 13 (fin) teak glazg. beads. | This is an instance where a dimensioned diagram, as prescribed in SMM L10.5.0.1.0 for screens is particularly appropriate, showing a single line outline of the framing with overall dimensions and sizes of component members. The screens are enumerated with details of the overall dimensions and including the framing and glazing beads by reference to the dimensioned diagram and the component drawing 10, from which the estimator can obtain all the relevant details for pricing. This avoids the need for lengthy descriptions. |
| 2/ | 3·37 | Bed wd. frs. in c.m. (1:3) & pt. o.s. in mastic. (sills | Bedding and pointing wood frames is measured in metres (SMM L10.10). |
| 2/ | 1·80 | (side (frames | |

```
            less      62        Glazg.
            rebs.2/13 26        width
                      36
                                 660
            less mulls. 2/½/36    36
                      75         624
                      13      ht.(max. sizes)
                      62     higher end  lower end
            less                2·400     1·900
            hd. 62
            sill 62             124       124
                               2·276     1·776

                                         2·276
                                         1·776
                                       2)4·052
                                 av. ht. 2·026
```

| | | | |
|---|---|---|---|
| 2/5/ | 0·62 2·03 | Glazg., stand. plain glass, panes area: 0·15 – 4·00 m²; pol. plate, 6th., to BS 952 to hwd. w. hwd. glazg. beads (m/s), secured w. brass cups & scrs. & beddg. in l.o. putty in irreg. shaped panes. | All the panes come within the 0·15 – 4·00 m² panes area classification, and so the areas of the different sized panes can be averaged in the dimensions. Raking cutting is deemed to be included with the item (SMM L40.C1) and so no labour is measured to the top edges. |

10·7

Painting

| | | | |
|---|---|---|---|
| 2/ | 1·80 | Paintg. gen. isoltd. surfs., hwd., gth. ≤ 300, appln. on site prior to fixg., primg. only. (side fr. to (masnry. | Painting required to be carried out on members on site before they are fixed shall be so described (SMM M60.1.0.2.4). The painting of glazed screens is measured each side in m², stating the panes area classification (SMM M60.2.4.1.0). However, the work to associated sills is measured separately as to general surfaces (SMM M60.M7). Hence the latter is taken as varnishing on general isolated surfaces, measured in metres as they are ≤ 300 mm girth, but it will be advisable to specifically mention this in the description of the glazed screens to avoid any possibility of duplication of pricing by the estimator. |

```
                    ht.
                    2·100
          less sill   75
                    2·025
```

| | | | |
|---|---|---|---|
| 2/ | 3·37 2·03 | Clear finishg., glazd. screens, hwd., panes area > 1·00 m², gth. > 300, 3 cts. of polyurethane clear varnish (varnishg. to sills m/s). & Ditto, ext. | The varnishing of the heads and uprights is included in the work on the screens. |
| 2/ | 3·37 | Clear finishg., gen. isoltd. surfs., hwd., gth. ≤ 300, 3 cts. of polyurethane clear varnish. (sills & Ditto, ext. | The work is deemed to be internal unless otherwise described (SMM M60.D1). |

10·8

# FOR READER'S NOTES

## DRG. 11

# FIXED BENCHING

**NOTES:**
All nosings, edgings, etc. to be in wrought Iroko
All sizes shown are finished sizes
All timber fixed to steel framing and supports by screws

### SECTION B-B
Scale 1:5

Callouts:
- 16⌀ bright chromium plate finish brackets and cable housing
- 38 x 28 hardwood nosing
- line of microphone in position
- fluorescent light fittings in chrome casing
- steel locating pin 9⌀ at 600 ccs
- welded joint
- 300 x 125 x 16g stainless steel fascia plate
- 12 thick plywood, hardwood veneered
- 38 x 6 m.s. flat framing
- removable bench service upstand of leather faced plywood 12 thick
- 30 x 25 hardwood nosing
- 25 chipboard, hardwood lipped and faced with leather
- service duct
- spacer and cable entry
- welded joint
- 20 thick plywood upstand hardwood veneered and leather faced
- 12 thick plywood hardwood veneered
- 16g aluminium microphone cable trunking
- 35 x 25 hardwood rail
- 54 x 25 hardwood edging
- 38 x 6 m.s. flat welded to upright
- 45 x 34 x 6 continuous m.s. angle welded to m.s. flat benchtop supports
- 18 x 65 hardwood edging
- 60⌀ 8 gauge m.s. tube upright with bright chromium plate finish
- 80 x 25 x 4 continuous m.s. channel welded to m.s. baseplate
- electric conduit
- 175 x 175 x 6 baseplate bolted to structural concrete at 400 ccs
- floor finish on screed
- welded connection
- 9⌀ x 100 long h.d. bolts Type RX3

### SECTION A-A

- end upright - flange on one side only
- quality 3S leather stuck to 6 plywood skin with adhesive
- 38 x 25 hardwood framing
- 25 x 3 m.s. flange welded to each side of upright (250 long)

### ELEVATION

- modesty panel

### END ELEVATION

- 18 french polished hardwood edging screwed to ends of framing with brass cups and screws

### PLAN   Scale 1:50

- removable sections

## FIXED BENCHING

| | | len. | |
|---|---|---|---|
| | | 3.230 | |
| | | 3.180 | |
| | | 6.410 | |
| | add ends 2/170 | 340 | |
| | len. of inner edge | 6.750 | |
| | add centre adjustment | 80 | |
| | len. on ℄ | 6.830 | |

| | | o/a ht | |
|---|---|---|---|
| | figd. ht. | 710 | |
| | add upstd. | 185 | |
| | | 895 | |
| | less edging | 65 | |
| | | 830 | |
| | add base sectn. below flr. fin. | 45 | |
| | o/a ht. | 875 | |

1     Metal bench dogleg base unit 6.83 m × 610 × 875 hi. o/a (fixg. to conc. flr. m/s), & tbr. dogleg benchtop 6.83 m × 610 mm o/a, in sel. wrot. Iroko protected. for subsqt. treatment & inc. leather facg. stainless stl. fascia plates & vert. modesty pans., all in accordance w. component details shown on Drg. 11.

The follg. in 1 nr. dogleg met. base unit 6.83 m × 610 × 875 hi.

| | ht. of ped. fittgs. | | |
|---|---|---|---|
| | figd. ht. | 710 | |
| | less chipbd. | 25 | |
| | | 685 | |
| | add flr. fin. | 12 | |
| | | 697 | |

11.1

---

This example illustrates one possible approach to measuring composite items.

The term 'composite item' refers to items to be manufactured off site as SMM General Rules 9.1. Fabrication in suitable sections for transport, installation and subsequent re-assembly is deemed to be included in the items.

It is considered desirable to give the overall dimensions of the fitting in the bill item and these are calculated in waste. The benching is likely to be fabricated off site and can thus be taken as an enumerated item containing suitable particulars and referring to the appropriate component details, in accordance with the rules for the measurement of fixtures (SMM N10.1.1-2.0.0).

The estimator will then need to build up his price from the details on the component drawing, taking off such quantities as he requires. One of the principal arguments for this approach was that the machine work in the fabricator's premises accounted for a significant proportion of the total cost and that listing the lengths of component members did not materially assist the estimator in determining the amount of labour involved, which he could more readily ascertain from detailed drawings.

Despite the lack of relevant guidelines in SMM7, it has been decided as an alternative approach to break the fitting down into its component parts as would be the case if the unit could not be fabricated off site.

---

## EXAMPLE 11

| | len. of bench suppt. | | |
|---|---|---|---|
| | o/a width | 610 | |
| | less upstand | 14 | |
| | edging | 18 | |
| | L | 34 | |
| | | | 66 |
| | | | 544 |

5     Pedestal fittg. 697 o/a ht. of 60∅ 8 gauge m.s. tube w. 2 nr 25 × 3 m.s. flanges 250 lg. welded to sides of tube the whole c.p. after manufacture; 38 × 6 m.s. benchtop suppt. 544 lg. welded to upper end of 60∅ m.s. tube; 38 × 6 flat m.s. vert. duct framg. 125 lg. welded to benchtop suppt.; 38 × 6 do. but slopg., 152 lg. & welded to top of vert. framg. & welded connectn of base of tube to m.s. chann. (m/s), w. all welds grd. smth.

2     Ditto. but w. 1 nr. 25 × 3 m.s. flange, welded to side of tube.

| | 6/800 | 4.800 | |
|---|---|---|---|
| | ends 2/300 | 600 | |
| | centre 2/620 | 1.240 | |
| | | 6.640 | |

11.2

---

This will give the student practice in identifying and scheduling the component parts and the major labours involved. It is emphasised that this approach is not intended in SMM7, but is nevertheless considered to be of benefit to the student, and difficulty has been experienced in locating a suitable non-prefabricated joinery item.

Figured dimensions are used as far as practicable and waste calculations are suitably headed and annotated, so that they may be readily followed and understood.

Particulars of the metalwork shall include the kind and quality of metal (normally included in a preamble clause), thickness or substance and method of fixing.

The measurement of the base unit commences with an enumerated item for pedestal fittings, comprising vertical tubes with flanges and associated benchtop supports and service duct framework, and including welded connections.

The end tubes have only one flange to support the modesty panel.

There is little guidance in SMM7 on the measurement of this class of work. The metalwork sections are restricted to L11 (metal windows/rooflights/screens/louvres), L21 (metal doors/shutters/hatches), L31 (metal stairs/walkways/balustrades), N10 (general fixtures/fittings/equipment) and P20 (unframed isolated trims/skirtings/sundry items).

104

| | | FIXED BENCHING (Contd.) | | | | |
|---|---|---|---|---|---|---|
| | 6·64 | M.s. chan., 80 x 25 x 4, welded to m.s. baseplate (m/s). | Bearers and the like are measured in metres, with a dimensioned overall cross-section description (SMM P.20.2.1.0.0). The method of jointing or form of construction and method of fixing should be stated where not at the discretion of the contractor (SMM P20.S7 & 8). The welded connections can be included in the description of the angle item. | | 3/800  2·400<br>centre  580<br>  300<br>  3·280<br>add tapd. end  130<br>  3·410 | It is not necessary to separately enumerate the irregular mitred angles on these members, as the sectional area does not exceed 0·003 m² and hence they are deemed to be included (SMM P20.C1). All timber sizes are deemed to be nominal sizes unless stated as finished sizes (SMM P20.D1). |
| | | len. of L to bench edge<br>  3·230<br>  3·180<br>  6·410 | | 3·45<br>3·41 | Edgg. 54 x 25 (fin.) spld. & tgd.<br>&<br>Nosg., 30 x 25 (fin.) spld. & grvd. | |
| | 6·41 | M.s. L, 45 x 34 x 6 to benchtop edge, welded to m.s. suppts. @ 800 ccs. | | | | |
| | | Baseplates<br>400 ) 6·830<br>17 + 1 = 18 | The number of baseplates is calculated in waste. | | scaled len.  6·750<br>add ends 2/60  120<br>  6·870 | |
| | 18 | M.s. baseplate, 175 x 175 x 6 ea. drilled & bolted to conc. w. 2 nr h.f. bolts, type RX3, 9 ϕ, 100 lg. | The baseplates are best enumerated with a dimensioned overall cross-section description including the fixing bolts. Formwork items to form mortices in concrete are enumerated and described as SMM E20.26. The same approach is now adopted for the timber benchtop to give the student some practice in identifying and measuring the component parts to what would normally be regarded as a composite item under SMM N10 and General Rules 9.1. The timber particulars follow the procedure prescribed for timber items generally. This edging is measured in the same way as a cover fillet, bead or nosing, in metres with a dimensioned overall cross-section description (SMM P20.2.1.0.0). The length of the 54 x 25 mm edging is calculated in waste from an accumulation of figured and scaled dimensions. The labours are included in the descriptions of linear items. | 6·87 | Rl., 35 x 25 (fin.) 2ce spld., 3ce grvd. & holl. grvd. | Note the description of the labours on this heavily worked member. The third groove is to receive the sloping mild steel framing. The length of each member varies and appropriate adjustments are necessary. The length of the rear face of the benchtop has been scaled in the absence of figured dimensions. The method of fixing is included in the component description where not at the discretion of the contractor (SMM P20.S8). The mean girth of the chipboard surface to the benchtop is calculated in waste. The width of the chipboard is not figured on the drawing, nor can it be scaled. Hence it is obtained by deducting the widths of the other components from the overall width. The particulars of the chipboard must include the type of sheet, thickness, method of jointing and method of fixing where not at the discretion of the contractor (SMM K13.S1-13), and protection for subsqt. treatment (leather facing). As it exceeds 300 mm wide, it is measured in m² (SMM K13). |
| | | The follg. in tbr. benchtop 6·83 m x 610 w/a scrd. to met. base unit (m/s), all in sel. wrot. Iroko, protected for subsqt. treatmt. & in accordance w. the component details shown on Drg. 11. | | 6·75 | Nosg., 38 x 28 (fin.) sply. reb, grvd. & 2ce pencil rdd., fxd. w. stl. locatg. pins, 9ϕ, 25 lg, @ 600 ccs. | |
| | | | | | chipbd.<br>len.<br>3·450<br>3·410<br>6·860<br>6·750<br>2 ) 13·610<br>av. len.  6·805<br>width<br>less  610<br>  upstd.  20<br>  slopg. stl. fascia  122<br>  tube  60<br>  pt. nosing  25<br>  edging  30<br>  outer edge (less reb.) 10  267<br>  343 | |
| | | Nosgs. & edgings | | | | |
| | 6·41 | Edgg., 18 x 65 (fin.) grvd., pencil rdd. on 2 edges. | | 6·81<br>0·34 | Chipbd. ling., width > 300, 25 th. to BS 2604 Pt. 2, hwd. lipped, reb., grvd. & glued to hwd. frmg. (m/s), protectd. for subsqt. treatmt. | |
| | | 54 x 25 edging<br>3/800  2·400<br>centre  620<br>  300<br>  3·320<br>add tapd. end  130<br>  3·450 | | | | |

11·3       11·4

FIXED BENCHING (Contd.)

| | | | | | |
|---|---|---|---|---|---|
| | | Plywood | | | |
| | | 6·870 | | | |
| | | 6·860 | | | |
| | | 2) 13·730 | | | |
| | | 6·865 | | | |

6·87    Plywood ling., width ≤ 300, 85 x 12 hwd. veneered, 2ce reb. & glued to hwd. frmg. (m/s).

The mean girth of the plywood section of the benchtop is calculated in waste, from the previous rail and nosing dimensions.

As the width does not exceed 300 mm wide, it is measured in metres (SMM K13). The jointing labours can conveniently be included in the descriptions of linear items.

6·86    Plywd. lining., width ≤ 300, 128 x 12 hwd. veneered, slopg., 2ce reb. & glued to hwd. frmg. (m/s).

Sheet linings laid sloping are so described (SMM K13). The width is scaled from the large scale drawing.

Raking cutting to plywood is deemed to be included (SMM K13.C1a).

6·75    Upstd. 20 x 170, plywd., hwd. veneered, spld. w. 2 pencil rdd. edges, & w. 7nr stopped mors.

This thicker member is measured under SMM P20.2.1.0.4, as it is in a different category to the plywood members measured previously. Stopped mortices to receive ends of benchtop supports are included in the component description.

1    E.O. hwd. veneered plywd. upstd. for mi.

Ends, angles, mitres and intersections are enumerated as the component is hardwood faced and the sectional area > 0·003 m² (SMM P20.C1).

2    Ditto. for end, spld.

It is assumed that the hardwood veneered members come within the classification of hardwood items.

| | Modesty Panels | |
|---|---|---|
| | la. panels | |
| | o/a len. | 1·500 |
| | less 2/½/60 | 60 |
| | | 1·440 |
| | sm. panels | |
| | o/a len. | 300 |
| | less 2/½/60 | 60 |
| | | 240 |

Calculation of the lengths of the modesty panels in waste (4 large panels and 2 smaller ones).

4    Panel, 1440 x 350, 2nr faces of plywd., 6th., ea. faced in leather, qual. 35, stuck w. adhesive, & glued to 38 x 25 (fin.) hwd. frmg. on all sides, w. holl. rdd. edges & grvs. (4nr stopped ends).

The panels are best enumerated with a dimensioned description as SMM P20.9.1.0.1 & 3. The size of the components and the number of stopped ends are included in the descriptions of the items. It seems realistic to include the leather facing with these items.

11·5

---

2    Ditto., 240 x 350.

| | Removable top sectns. | |
|---|---|---|
| | | gth. |
| | | 38 |
| | | 145 |
| | | 120 |
| | | 303 |

4    E.O. benchtop for formg. removable sectn., 1500 x 303.

The forming of removable top sections are taken as enumerated 'extra over' items, stating the size, to enable the estimator to cover the extra costs involved.

Ends to benchtop

2/1    Edgg. wrot hwd., 340 x 65 x 18 th., fxd. to hwd. frmg. (m/s) w. brass cups & screws.

In view of the relatively short lengths involved, it is considered preferable to enumerate these items, rather than to take them in metres, in accordance with SMM P20.2.1.0.0.

2/1    Edgg. wrot hwd., 340 x 185 (extreme depth) x 18th., L shaped sectn. as component detail (Drg. 11), fxd. to hwd. frmg. (m/s) w. brass cups & screws.

Because of the irregular shape, this item is best covered by a dimensioned diagram or reference to a component detail. The irregular mitred intersection will not form a separate enumerated item, as the cross sectional area of the edging is less than 0·003 m² (SMM P20.C1).

2/1    Clear finishg., gen. isoltd. surfs, hwd., area ≤ 0·50 m²., & 2ce French pol.

The polishing to the edging to each end of the benchtop is taken as an enumerated item in accordance with SMM M60.1.0.3.0, as ≤ 0·50 m² in area. The work is classified as to general isolated areas.

Steel plates

8    Fascia plate, stainless steel, 300 x 125 x 16g, scrd. to plywd. (m/s) w. c/sk. stl. screws.

This is an enumerated item in accordance with SMM P20.9.1.0.0, with a dimensioned description and stating the method of fixing where not at the discretion of the contractor (SMM P20.S8).

| Leather facg. | |
|---|---|
| Girths | |
| Upstd. | |
| spld. top | 20 |
| outer vert. face | 157 |
| bott. edge | 20 |
| inner vert. face | 37 |
| | 234 |

The girths of leather facing to joinery components are calculated in waste, including the covering to edges.

11·6

| | | | | |
|---|---|---|---|---|
| | | FIXED BENCHING (Contd.) | | |
| | | Fascia & part benchtop | | The lengths are taken from those previously calculated for the component members to the benchtop. |
| | | top nosing | 18 | |
| | | " " | 27 | |
| | | " " | 30 | |
| | | " " | 15 | |
| | | plywd. fascia | 118 | |
| | | rail | 22 | |
| | | plywd. panel | 78 | |
| | | 30 x 25 nosg. | 30 | |
| | | " " | 30 | |
| | | | 368 | |
| | | Inner surf. of benchtop | | |
| | | 54 x 25 edging | 30 | |
| | | " " | 30 | |
| | | chipbd. | 333 | |
| | | outer edging, edges 2/18 | 36 | |
| | | o/edg. outer face | 65 | |
| | | o/edg. inner face | 44 | |
| | | | 538 | |
| 6.75 | | Leather coverg. quality 35 stuck to hwd. w. apprvd. adhesive. | | The only guidance for the measurement of this class of work is contained in SMM M50 and even this is not altogether appropriate. It is therefore measured in m², giving a description of the work, quality of materials and nature of the base to which the leather is applied as SMM M50. S1 & 5. The work is deemed to include fair joints, working over and around obstructions, into recesses and shaped inserts, and fixing at the perimeter (SMM M50.C1).
The microphones and fluorescent light fittings will probably be covered by prime cost items. |
| 0.23 | | | | |
| 6.86 | | | | |
| 0.37 | | | | |
| 6.81 | | | | |
| 0.54 | | | | |

# 7 FINISHINGS

## SEQUENCE OF MEASUREMENT

It is important to adopt a logical sequence of measurement and a common approach is to work a floor at a time taking the order of ceilings, walls and floors. The measurement of the areas of wall finishings will normally be followed by linear items such as cornices, coves, picture rails, dado rails and skirtings, and working in this sequence (top to bottom).

Practices will, however, vary from office to office and floor finishings, for example, could be taken with the floors or left later to be measured with the finishings. On balance it is probably better to take boarded flooring with the floor construction as it forms an integral part of this element of the structure. Finishings to solid floors, however, can be of infinite variety, as shown in example 12, and together with the appropriate beds are best measured with the finishings.

A schedule of finishings of the type prepared for example 12 may be supplied by the architect and, if not, the quantity surveyor is advised to prepare his own in order that the various finishings are systematically recorded to give the overall picture and to identify the extent of similar finishings which are usually grouped together when taking off to avoid the duplication of similar items. Rooms should be named or numbered for identification purposes and the locations inserted in waste when taking off so that the locations of measured items are evident. As each finishing item is measured the schedule of finishings can be suitably crossed through to reduce the risk of omissions.

## GENERAL PRINCIPLES OF MEASUREMENT

The surface finishes or finishings section of the bill will refer to the location drawings or any further drawings which accompany the bill of quantities for the location and scope of the work (SMM M10/20/30/40/50.P1a). The work is deemed to be internal unless described as external (SMM M20.D1), to enable the estimator to make allowance for the different working conditions in his prices. Where finishings are applied to confined or awkward spaces (in staircases and plant rooms), this work is measured and described separately in each case (SMM M20.M3), and will usually generate a higher price rate. Other work that could prove to be more expensive, such as multi-coloured painting, also requires separate itemising.

The nature of the base to which the finishings are to be applied is stated, and similar finishings applied to different bases, such as concrete, brickwork and blockwork, are kept separate. Hacking concrete to form a key for a surface finish constitutes a separate item (SMM E41.4.0.0.0), whereas raking out joints of brickwork, blockwork or stone walling are deemed to be included (SMM F10/F21.C1d). Curved work is also separated because of its higher cost, as illustrated in example 12, and so described, stating the radii measured on face (SMM M20.M5).

### Ceilings

The areas of ceilings are measured between wall surfaces, with the area of each type of finish measured in $m^2$ where the width > 300 mm, otherwise in metres, followed by associated labours such as arrises to beams. Plastered ceilings are to include the relevant particulars listed in SMM M20.S1–6, such as the kind, quality, composition and mix of materials, method of application, nature of surface treatment and nature of base. While plasterboard ceilings are dealt with similarly, stating the thickness, and are deemed to include joint reinforcing scrim (SMM M20.C3).

Work to ceilings and beams over 3.50 m above floor (measured to ceiling level in both cases), except in staircase areas, are kept separate and classified in further stages of 1.50 m, that is, >3.50 m and ≤ 5.00 m, > 5.00m and ≤ 6.50 m and so on (SMM M20.M4). Work to the sides and soffits of attached beams and openings is classified as work to abutting ceilings (SMM M20.D5). Work in staircase areas is given separately to enable the estimator to price for work off staircase flights and/or in restricted areas (SMM M20.M3).

### Walls

The measurement of wall finishes is normally taken from floor to ceiling, including the work behind wood skirtings and similar features, disregarding the grounds (SMM M20.M2).

The girth of each room is usually built up in waste and it is often good practice to transfer the total girth of rooms of the same height and finish to the dimension column. The method of measurement varies with the type of finish and reference needs to be made to the appropriate work section, such as M20 for renderings and plastered coatings, M40 for quarry and ceramic tiling and M41 for terrazzo tiling. The measurement of the areas of wall finish of each type is followed by the associated linear items.

Plaster to walls and isolated columns in widths $\leq 300$ mm is measured in metres and greater widths are taken in $m^2$. No deductions are made for voids $\leq 0.50\ m^2$ (SMM M20.M2), but it must be remembered that openings or wants which are at the boundaries of measured areas are always deducted irrespective of size (SMM General Rules 3.4).

Work to sides and soffits of openings and sides of attached columns shall be regarded as work to the abutting walls. Rounded angles and intersections to plaster in the range 10–100 mm radius are measured in metres (SMM M20.16 and M20.M7), while those of smaller radius are included in the plasterwork rates.

Working plaster over and around obstructions, pipes and the like, and into recesses and shaped inserts is deemed to be included (SMM M20.C1b).

Cornices, mouldings and coves are measured in metres (length in contact with base) stating the girth or giving a dimensioned description (SMM M20.17–19.0.1–2.0). Ends, internal angles, external angles and intersections are each enumerated extra over the appropriate linear items, giving adequate details as SMM M20.23.1–4.1.0.

Where the wall finish is of tiles, slabs or blocks of cast concrete, precast terrazzo, natural stone, quarry or ceramic tiles, plastics, mosaic or cork, full particulars of the finishes are to be given, such as the kind, quality and size of materials, nature of base, surface treatment, method of fixing, and treatment and layout of joints (SMM M40/41/50.S1–8). Work > 300mm in width is measured in $m^2$, while narrower widths are measured in metres.

*In situ* finishes are measured in a similar manner to plastering.

Example 12 illustrates the method of measurement of several different types of wall finish.

### Floors

Floor finishes are measured in $m^2$, irrespective of their width and are classified under three categories according to slope (SMM M40.5.1–3), and where floors are laid in bays, the average size of bay is stated. It is customary with solid floors to vary the thickness of the bed or screed, according to the thickness of the finish, to give a uniform overall thickness and a level floor surface. Descriptions of screeds are to include the kind, quality, composition and mix of materials, method of application and nature of base (SMM M10.S1, 2 and 5).

Areas of floor finishes in door openings are taken with the adjoining floor finishes. The boundary between different floor finishes normally occurs at the side of the dividing wall from which the door is hung, so that the joint between the floor finishes is hidden when the door is shut. Dividing strips at door openings are measured in metres, giving a dimensioned description as SMM M40.16.4.1.0.

Tile and block floors, such as stone, concrete, quarry, ceramic, mosaic, wood block, composition block and parquet flooring, may be classified as plain or as work with joints laid out to detail, including a dimensioned diagram, and may also include details of patterned work, the average size of bays where floors are laid in bays, and size or section of inserts (SMM M40/42.5.1–3.1–2.1–3). The particulars of the floor are to include details of the materials, size, shape and thickness of units, nature of base, preparatory work, nature of finished surface including any sealing/polishing, bedding or other method of fixing, and treatment and layout of joints (SMM M40/42.S1–8).

The descriptions of rubber, plastics, cork, lino and carpet tiling and sheeting, and edge fixed carpeting shall contain the relevant particulars listed in SMM M50/51.S1–8, such as the kind, quality and size of materials, nature and number of underlays, extent of laps, type of seams, nature of base, surface treatment, pattern, width and laying direction of materials, and method of fixing and treatment of joints. Movement joints and cover strips are measured in metres with a dimensioned description (SMM M50/51.13.3–4.1.0).

## OTHER FINISHINGS

Sprayed mineral fibre coatings are measured in m², stating the thickness and number of coats or the thickness of plasterboard or other rigid sheet lathing and thickness and number of coats, categorising them according to location and distinguishing between walls and columns, ceilings and beams, and structural metalwork (SMM M22.1–3.0.1–2.0).

Metal mesh lathing and anchored reinforcement for plastered coating are classified as to location, giving the depth of suspension in three categories for suspended lathing ceilings and the method of fixing to the structure (SMM M30.1–5.1–3.1.0). The work is deemed to include mechanical fixings, steel channel framing, ancillary fixing materials, additional support and trimming for light fittings and internal and external angles < 100 mm radius (SMM M30.C1).

Fibrous plaster is measured in m² where the width > 300 mm and in metres for narrower widths, stating the thickness, and giving details of the materials, method of fixing, treatment of joints, nature of base, and details of timber or metal lathing and reinforcement, and distinguishing between work to walls and ceilings (SMM M31.1–2.1–2.1.0 and M31.S1–4). Coves, mouldings, cornices and architraves (measured the extreme lengths) are each given separately in metres with a dimensioned description (SMM M31.10–13.0.1.1–5), while ends, internal angles, external angles and intersections are each enumerated and detailed as extra over the work in which they occur (SMM M31.14.1–4.1.0).

## SKIRTINGS AND SIMILAR MEMBERS

Timber skirtings, picture rails, dado rails, architraves and the like are measured in metres, giving a dimensioned overall cross-section description as SMM P20.1.1.0.1–4. Ends, angles, mitres, intersections and the like are deemed to be included in these items, except where the members are in hardwood > 0.003 m² sectional area, in which case they are enumerated (SMM P20.8.1–4.0.0). The description is to include the kind and quality of timber, whether sawn or wrought and the method of fixing where not at the discretion of the contractor (SMM P20.S1–9).

*In situ* and tile, slab and block skirtings are measured in metres, stating the height or height and width as appropriate, and are deemed to include fair edges, rounded edges, ends, angles and ramps (SMM M40.12.1–2.0.0 and M40.C8). It will be seen in example 12 that the measurement of the coved *in situ* terrazzo skirting involves the deduction of areas of wall lining and floor paving, as the skirting is not measured extra over these items (SMM M22.1–3.0.1–2.0).

## PAINTING AND DECORATING

The painting and decorating of ceilings, cornices and walls are classified as to general surfaces and measured in m², except for work on isolated surfaces ≤ 300 mm in girth which is given in metres, or work in isolated areas ≤ 0.50 m² which is enumerated (SMM M60.1.0.1–3.0). Multi-coloured work is separately classified and is defined as the application of more than one colour on an individual surface, except on walls and piers or on ceilings and beams (SMM M60.D2). Paintwork is deemed to include rubbing down with glass, emery or sand paper (SMM M60.C1). Full particulars of painting, decorating and polishing shall be given in accordance with SMM M60.S1–8.

Work in staircase areas and plant rooms is to be kept separate because of the extra costs involved (SMM M60.M1). Work to ceilings and beams over 3.50 m above floor level (measured to ceiling level in both cases), except in staircase areas, shall be so described stating the height in further 1.50 m stages (SMM M60.M4).

The supply and hanging of decorative papers and fabrics is separated between walls and columns; and ceilings and beams; with areas > 0.50 m² measured in m² and those ≤ 0.50 m² enumerated. Where these items include raking and curved cutting and/or lining paper, these are inserted in the description (SMM M52.1–2.1–2.0.1–2). Border strips are measured in metres, including cutting them to profile in the description where appropriate (SMM M52.3.0.0.1). Material particulars are to include the kind and quality of materials, including the manufacturer and pattern (SMM M52.S1), although these particulars may be included in a preamble clause or be covered by a project specification reference. The *Code of Procedure for Measurement of Building Works* states that the width of rolls and type of pattern would need to be given before wallpaper could be considered fully described.

DRG. 12                                         FINISHINGS

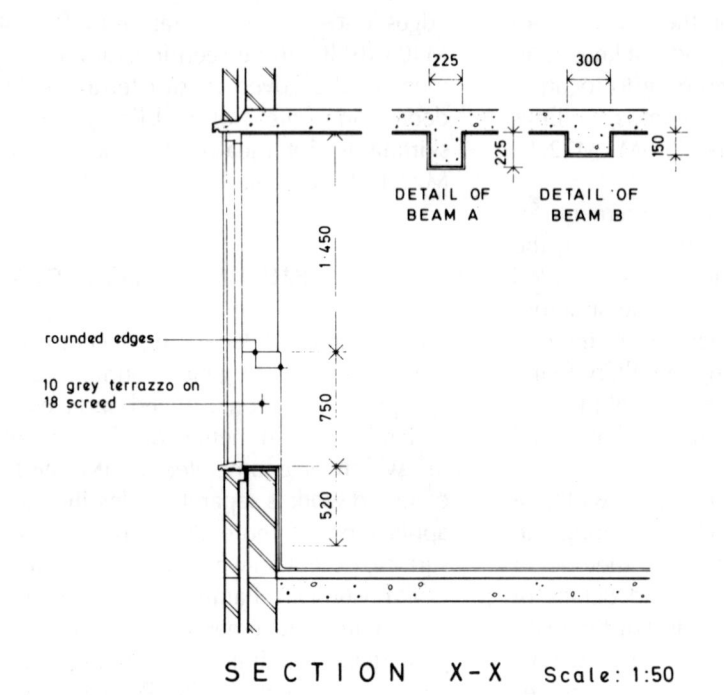

**SECTION X-X**  Scale: 1:50

- rounded edges
- 10 grey terrazzo on 18 screed
- 1·450
- 750
- 520
- 225
- 300
- 150
- DETAIL OF BEAM A
- DETAIL OF BEAM B

Note:
See details of finishings to ceilings, walls and floors on accompanying schedule

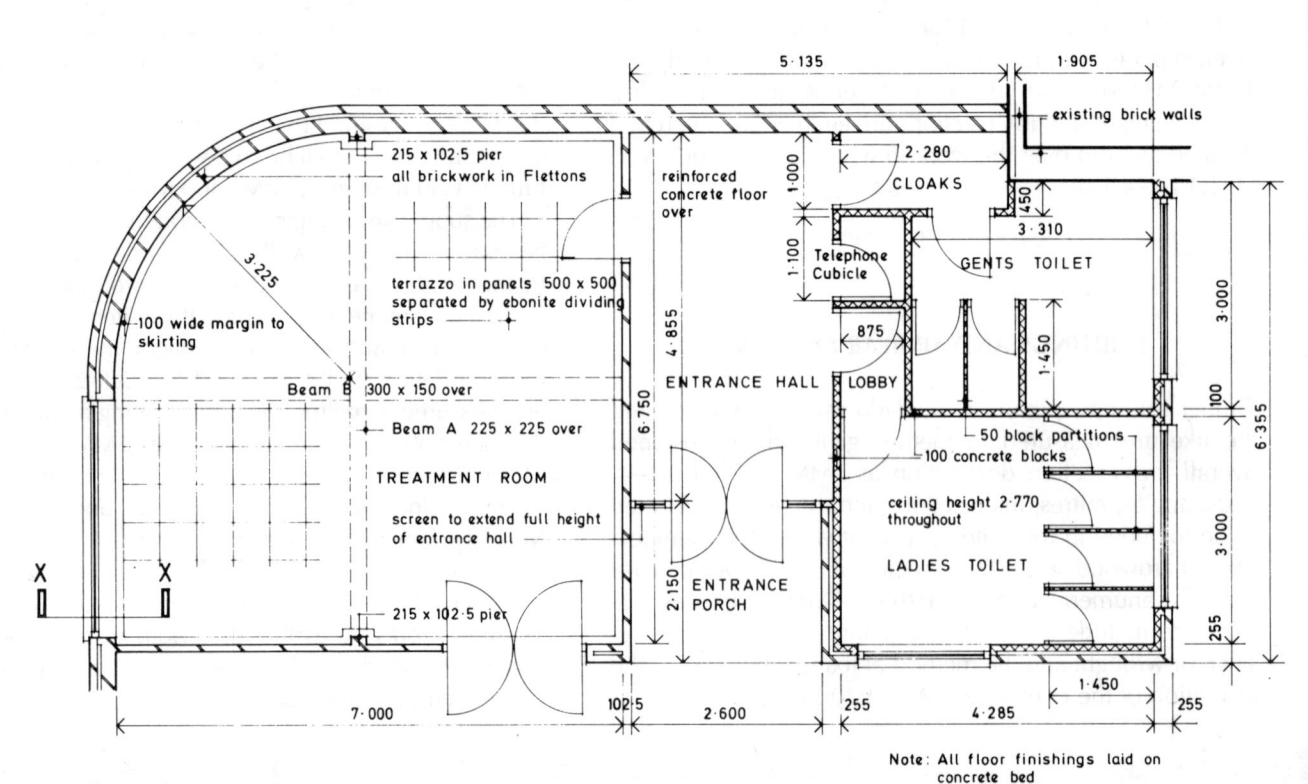

**PLAN**  Scale 1:100

- 215 x 102·5 pier
- all brickwork in Flettons
- reinforced concrete floor over
- existing brick walls
- CLOAKS
- Telephone Cubicle
- GENTS TOILET
- terrazzo in panels 500 x 500 separated by ebonite dividing strips
- 100 wide margin to skirting
- Beam B 300 x 150 over
- Beam A 225 x 225 over
- TREATMENT ROOM
- screen to extend full height of entrance hall
- 215 x 102·5 pier
- ENTRANCE HALL LOBBY
- 50 block partitions
- 100 concrete blocks
- ceiling height 2·770 throughout
- ENTRANCE PORCH
- LADIES TOILET

Dimensions: 5·135, 1·905, 2·280, 1·000, 1·450, 3·310, 1·100, 875, 4·855, 6·750, 3·225, 1·450, 3·000, 100, 6·355, 3·000, 255, 7·000, 102·5, 2·600, 255, 4·285, 1·450, 255, 2·150

Note: All floor finishings laid on concrete bed

## SCHEDULE OF FINISHINGS

| Location | Ceiling Finishing | Decorations to Ceilings | Wall Finishing | Decorations to Walls | Skirting | Floor Finishing | Any Other Features |
|---|---|---|---|---|---|---|---|
| Treatment Room | 10 thick, 2 coat lightweight gypsum plaster to BS 1911 Part 2 on concrete with bonding plaster backing and final coat of finish plaster steel trowelled | Prepare and apply 3 coats of white emulsion | 10 thick grey *in situ* terrazzo dado, 1.27 m high above skirting on 12 thick cement and sand (1:4) screeded backing plaster above as for ceilings | Prime and apply 3 coats of oil paint, matt finish, above dado | 10 thick black *in situ* terrazzo 150 high flush with dado above and with terrazzo margin 16 thick and 100 wide at junction with floor | 16 thick black and white *in situ* terrazzo laid alternately in panels about 500 × 500 between 6 – 16 ebonite dividing strips on building paper underlay on 34 thick cement and sand (1:3) screeded bed | Rounded edge to wall terrazzo at top edge and rounded external and coved internal angles. 10 radius coved internal vertical angles and rounded arrises to plasterwork in all compartments |
| Entrance Porch | 25 thick British Columbian Pine tongued, grooved and V jointed boarding nailed to 50 × 19 impregnated softwood battens plugged to concrete | K.p.s. and 3 coats of oil paint gloss finish | Cement and sand (1:4) backing 10 thick and Tyrolean finish of 'Cullamix' mixture applied by machine | — | — | 152 × 152 × 22 brown quarry tiles to BS 1286 Type A, bedded, jointed and pointed in cement mortar (1:3) on 28 thick cement and sand (1:3) screeded bed | |
| Entrance Hall | 10 thick 2 coat lightweight gypsum plaster as Treatment Room | Prime and apply 3 coats of oil paint, matt finish | 4 thick pre-finished Iroko veneered plywood decorative panelling in sheets 2.44 × 1.22 m with random V groove on face to dado 1.22 m high above skirting, fixed with adhesive to 50 × 19 impregnated softwood battens fixed to walls at 406 centres and top and bottom. 13 thick 2 coat plaster above dado with base coat of Thistle browning and sand (1:2) 11 thick and finishing coat of Thistle finish plaster 2 thick steel trowelled | Prime and apply 3 coats of oil paint, matt finish above dado | 38 × 225 Iroko chamfered skirting, on grounds and apply 2 coats of polyurethane clear varnish | 25 thick Iroko tongued and grooved wood block flooring laid herringbone pattern bedded in mastic with 2 block wide border and machine sanding on 25 thick cement and sand (1:3) bed. Seal, body in and twice wax polish | 32 × 32 moulded Iroko rail to top of plywood dado. Cornice of gypsum plaster (class B) to BS 1191 Part 1, with curved contour of 125 girth |
| Cloaks | 10 thick, 2 coat lightweight gypsum plaster as Treatment Room | Prepare and apply 2 coats of emulsion paint | 13 thick, 2 coat gypsum plaster to BS 1191 Part 1 of browning plaster and sand backing and final coat of Class B finish plaster, steel trowelled | Prepare and apply 2 coats of emulsion paint | 25 × 150 chamfered softwood skirting on grounds, and k.p.s. and 3 coats of oil gloss finish | 300 × 300 × 2 PVC tiles to BS 3261 Type A fixed with adhesive on 48 thick cement and sand (1:3) trowelled bed | |
| Telephone Cubicle | Fibreboard accoustic tiles 305 × 305 × 19 butt jointed both ways, fixed with adhesive to cement and sand (1:4) backing, 10 thick | — | 13 thick, 2 coat gypsum plaster as Cloaks | Prepare and apply 2 coats of emulsion paint | 25 × 150 chamfered softwood skirting on grounds, and k.p.s and 3 coats of oil gloss finish | 300 × 300 × 3.2 cork tiles, Cork-o-Plast, checker tiles Nr 13 standard, semi bright finish, butt jointed, fixed with copad adhesive on 47 thick cement and sand (1:3) trowelled bed | |
| Ladies and Gents Toilets and Lobby | 10 thick, 2 coat plaster with 8 thick coat of Carlite bonding plaster finished with 2 thick coat of Carlite finish plaster | Prime and apply 3 coats of oil paint, gloss finish | 152 × 152 × 6 cushion edged glazed ceramic tiles to BS 1281, fixed with adhesive jointed and pointed with white cement on 10 thick, cement and sand (1:4) floated backing for full height above skirting | — | — | 152 × 152 × 9.5 full vitrified ceramic tiles to BS 1286, with 3 wide joints, bedded, jointed and pointed with cement mortar (1:3) on 40 thick cement and sand (1:4) screeded bed | External angles to wall tiling to be finished with rounded edge tiles |

## FINISHINGS

|  |  |  |  |
|---|---|---|---|
|  |  | Ceilgs. Treatment Rm. | |
|  |  | 6·750 | |
|  |  | less beam 300 | |
|  |  | 2) 6·450 | |
|  |  | 3·225 | |
|  |  | less 3·225   7·000 | |
|  |  |       225   3·450 | |
|  |  |             3·550 | |

| 2/ | 3·55 |
|---|---|
|    | 3·23 |

Pla. clg., width > 300, lightwt. gypsum pla. to BS 11911 Pt. 2 in 2 cts, 10 th. %a. on conc. w. bondg. pla. backg. & final ct. of fin. pla. trowld. smth.

&

|    | 3·23 |
|---|---|
|    | 3·23 |
| 1/4/22/7 | 3·23 |
|    | 3·23 |

Paintg. gen. surfs. gth > 300, seal & 3 cts. white emulsn. pt., pla. clg.

A logical sequence is adopted for the measurement of the finishings, taking the order of ceilings, walls, skirtings and floors.
The work is deemed to be internal unless described as external (SMM M20.D1), and so internal work requires no specific mention.
The areas of ceiling to the Treatment Room between the beams are calculated in waste. The plasterwork to the beams has to be kept separate as the widths of the sides and soffits are ≤ 300 mm and are measured in metres (SMM M20.2.1-2.1.0) but classified as to ceilings (SMM M20.D5). The description of the ceiling plaster includes the thickness and number of coats and the relevant particulars listed in SMM M20.S1-6, including the kind, quality, composition and mix of materials, nature of base, preparatory work and surface treatment.
One section of the ceiling is in the form of a quadrant ($\frac{1}{4}\pi r^2$). Decorations to ceilings are measured in m² and classified as to general surfaces (SMM M60.1.0.1.0), with particulars given in accordance with SMM M60.S1-8. Painting and decorations to ceilings > 3·50 m above floor level are so described in 1·50 m stages, except in staircase areas which are kept separate and so described (SMM M60. M1 & M4). Rubbing down is deemed to be included and does not require specific mention (SMM M60.C1).

|  |  | beam B |
|---|---|---|
|  |  | 3·550 |
|  |  | 3·225 |
|  |  | 6·775 |

12·1

## EXAMPLE 12

| 3/ | 6·75 | Pla. clg., width ≤ 300, lightwt. gyp. pla. 10 th. %a on conc. a.b.d. | (beam A soff. & sides |
|---|---|---|---|
|    | 6·78 |    | (beam B soff. |
| 2/ | 6·78 |    | (beam B sides |

The plasterwork to the beams is classified as to ceilings, but as the width is ≤ 300 mm, it is measured in metres (SMM M20.2.2.1.0 and M20.D5).

| 3/ | 6·75 | Paintg. gen. surfs., gth. > 300, seal & 3 cts. white emulsn. pt., pla. clg. | (beam A soff. & sides |
|---|---|---|---|
|    | 0·23 |    |   |
|    | 6·78 |    | (beam B soff. |
|    | 0·30 |    |   |
| 2/ | 6·78 |    | (beam B sides |
|    | 0·15 |    |   |

The decorations to the sides and soffits of the beams are added to similar work on the ceiling, where it is of the same colour, and are measured in m² as they do not form isolated surfaces.

| 2/ | 0·10 | Ddt. pla. clg., width ≤ 300, a.b. | (piers |
|---|---|---|---|
| 2/ | 0·22 | Ddt. paintg. gen. surfs., gth. > 300, emulsn. pt. pla. clg. a.b. | (piers |
|    | 0·10 |    |   |

|  |  | 6·750 |
|---|---|---|
|  | less ends over piers 2/102·5 | 205 |
|  |  | 6·545 |

| 2/ | 6·55 | Rdd. L to pla., rad. 10 - 100. | (beam A |
|---|---|---|---|
| 2/ | 6·78 |    | (beam B |

Both plasterwork and painting are deducted to the soffit of beam B where it is supported by piers, as these areas are at the boundary of the measured area (SMM General Rules 3.4). No deduction is made at the intersection of the beams as the areas involved are less than the void limit of 0·50 m² (SMM M20.M2 and M60.M3).
Rounded angles are measured in metres when in the range 10 - 100 mm radius (SMM M20.16.0.0.0 and M20.M7). The work is deemed internal unless described as external (SMM K11.D1 and M60.D1).
Boarding to the entrance porch ceiling is measured in m² with a dimensioned description and stating the width range, location, method of jointing and fixing where not at the discretion of the contractor and nature of background as SMM K11.3.1.1.0 and K11.S1-13. The painting particulars are given in accordance with SMM M60.1.0.1.0 and M60.S1-8.

|  |  | Ent. Porch |
|---|---|---|
|  |  | 2·150 |
|  | less side lts. | 100 |
|  |  | 2·050 |

|  | 2·60 | Tbr. bd. lin̂g. to clg. width > 300, 25 th. wrot. Brit. Col. Pine t.&g. & V. jtd. bdg. nailed to swd. battens (m/s), ext. |
|---|---|---|
|  | 2·05 |   |

&

Paintg. g.s., gth. > 300, k.p.s., ② u/cs. oil paint & ① gloss fin. to wd. clg., ext.

12·2

| | | | |
|---|---|---|---|
| FINISHINGS | (Contd.) | | |

```
                400) 2·600
                     7+1
```

| | |
|---|---|
| 8/ 2·05 | Individl. suppts., 50 x 19 impregntd. sawn swd. plugd. to conc. |

Battens are measured in metres as individual supports with a dimensioned overall cross-section description (SMM G20.13.0.1.0), which includes the method of fixing where not at the discretion of the contractor (SMM G20. S2).

Continue to work round the building in a logical sequence marking off each section on the Schedule of Finishings, as it is completed. Make full use of waste for the build up of dimensions.

```
               Ent. Hall
                       2·600
         add cavity wall 255
                       2·855
         less block ptn. 100
                       2·755
```

| | |
|---|---|
| 4·86<br>2·76 | Pla. clg., width > 300, ltwt. gyp. pla. to BS 1192 Pt.2 in 2 cts., 10 th. o/a on conc.<br>&<br>Paintg. g.s., gth. > 300, ① primg. ct., ② u/cs oil pt. & ① matt fin. to pla. clg. |

The description of the ceiling plaster is reduced by the use of the letters 'a.b.d.' (as before described) as it consists of the same materials as used for the Treatment Room.

For painting give particulars of the kind and quality of materials, painting requirements and nature of base (SMM M60.S1-8).

```
               Cloaks
```

| | |
|---|---|
| 2·28<br>1·00 | Pla. clg., width > 300, ltwt. gyp. pla. to BS 1911 Pt 2 in 2 cts., 10 th. o/a on conc. a.b.d.<br>&<br>Paintg. gen. surfs., gth. > 300, seal & 2 cts. of emulsn. pt., pla. clg. |

It appears from SMM M20.2.1.1.0 and the SMM7 Library of Standard Descriptions that where the width > 300 mm, this is to be stated in the description. This does, however, seem superfluous as it could be assumed to be > 300 mm wide, if not described as ≤ 300 mm and measured in metres.

Decorations are described in accordance with SMM M60.1.0.1.0 and M60.S1-8.

The description of the acoustic tiles includes the kind, quality and sizes of tiles, nature of base, surface treatment, method of fixing and treatment of joints (SMM M50.S1-8). The backing forms a separate item, to which the same rules apply as for cement sand screeds (SMM M10.2.1.1.0).

```
          Telephone Cubicle
```

| | |
|---|---|
| 1·10<br>0·88 | Fibrebd. accoustic self-decorated tiles to clg., width > 300, 305 x 305 x 19 th. butt jtd. to symmetrical layout, fxd. w. adhesive to ct. & sd. backg. (m/s).<br>&<br>Ct. & sd. (1:4) backg. to clg., width > 300, stl. trowelled 10 th., on conc. |

12·3

```
         Ladies & Gents
         Toilets & Lobby
               Lobby
                 1·450
     less blk. ptn. 100
                 1·350

            Gents Toilet
                 3·000
     less recess    450
                 2·550
```

| | |
|---|---|
| 3·31<br>2·55 | Pla. clg., width > 300, in 2 cts., 10 th. (gents toilet) o/a w. 8 th. ct. of Carlite bondg. finish & 2 th. ct. of Carlite (ditto recess) fin. pla. to conc.<br>& |
| 1·91<br>0·45 | |
| 4·29<br>3·00 | Paintg., g.s., gth. > 300, ① primg. ct., (ladies toilet) ② u/cs oil paint & (lobby) ① gloss fin., pla. clg. |
| 1·35<br>0·88 | |

In building up areas, figured dimensions are used wherever possible in preference to scaling.

Another alternative type of ceiling plaster is used to give the student examples of different approaches and increased experience in framing descriptions. The locations of the areas are shown in waste to help in identifying dimensions in the future.

The sectional area of each dividing partition in the toilet cubicles is ≤ the void limit of 0·50 m² prescribed in SMM M20.M2 and M60.M3, and hence they do not have to be deducted from the area of ceiling.

Note use of different abbreviations of 'gen. surfs.' and 'g.s.' for general surfaces. The girth of straight lengths of walling, excluding piers, is calculated in waste.

```
               Walls
          Treatment Rm.
                    6·750
      2/3·550       7·100
      2/3·225       6·450
         beam        300
                   20·600
```

| | |
|---|---|
| 20·60<br>1·27 | Terrazzo in situ ling. to walls, width > 300, 10 th., of ct. & grey marble chippgs. (1:2), trowelled to smth. fin. & pol. on ct. & sd. backg. (m/s) |

```
                   1·270
         add sktg.  150
                   1·420
```

| | |
|---|---|
| 20·60<br>1·42 | Ct. & sd. (1:4) backg. to walls, width > 300, fltd., 12 th., on bwk. |

The terrazzo dado description includes the appropriate particulars contained in SMM M10.1.1.1.0. This is followed by the cement and sand backing with a similar approach but the height is increased by 150 mm to include the skirting.

12·4

113

# FINISHINGS (Contd.)

| | | |
|---|---|---|
| 1/4/22/7 | 3.23<br>1.27 | Terrazzo in situ ling. to walls, width > 300, 10 th. a.b.d. curved to rad. of 3.23 m. |
| 1/4/22/7 | 3.23<br>1.42 | Ct. & sd. (1:4) backg. to walls, width > 300, fltd., 12 th., on bwk., curved to rad. of 3.23 m. |
| 2/<br>2/2/ | 1.27<br>1.27 | Terrazzo in situ ling. to walls, width ≤ 300, 10 th., a.b.d. (pier faces) (pier retns.) |
| 2/<br>2/2/ | 1.42<br>1.42 | Ct. & sd. (1:4) backing to walls, width ≤ 300, 12 th. on bwk. (pier faces) (pier retns.) |
| | 20.60 | Rdd. L to terrazzo, rad. 10-100. (top edge) |
| 1/4/22/7 | 1.27 | (piers) |
| 1/4/22/7 | 3.23 | Ditto. curved to rad. of 3.23 m. (top edge) |
| 3/<br>2/2/ | 1.27<br>1.27 | Rdd. L to terrazzo, rad. 10-100. (corners to rm.) (piers) |
| | 20.60<br>1.45 | Pla. to walls, width > 300, lightwt. gyp. pla. to BS 1911 Pt. 2 in 2cts., 10 th. o/a to bwk. w. bondg. pla. backing & final ct. of fin. pla., trowld. fin.<br>&<br>Paintg. g.s., girth > 300, ① primg. ct., ② u/cs oil paint & ① matt. fin. to pla. walls. |

Curved work is taken separately, stating the radius measured on face (SMM M10. M5).

In situ finishings to areas ≤ 300 mm wide are measured in metres as SMM M10.1.2.1.0.

Rounded angles are measured in metres when in 10-100 mm radius range as SMM M10.16. 0.0.0 and M10.M7. Curved work is given separately stating the radius (SMM M10. M5). If the rounded angles were ≤ 10 mm radius, they would not be measured; where > 100 mm radius they are classified as curved work (SMM M10. D3).

The plaster above the terrazzo dado is described giving the appropriate particulars from SMM M20.1.1.1.0 and M20. S1-8. The height is taken from Section X-X. Measurements are taken on the actual wall surfaces (area in contact with base); not on the centre line of the plaster (SMM M20.M2). The paintwork description includes the preparation, kind of paint, number of coats and nature of base (SMM M60. S1-8).

12.5

| | | |
|---|---|---|
| 2/ | 0.30<br>0.15 | Ddt both last (ends of 300 × 150 beam) |
| 1/4/22/7 | 3.23<br>1.45 | Pla. to walls, width > 300, lightwt. gyp. pla. a.b.d. to curved surf. to rad of 3.23 m.<br>&<br>Paintg. g.s., gth. > 300, ① primg. ct., ② u/cs. oil paint & ① matt. fin. to pla. walls. |
| | | less beam 1.450<br>               225<br>               1.225 |
| 2/<br>2/2/ | 1.23<br>1.23 | Pla. to walls, width ≤ 300 lightwt. gyp. pla. a.b.d. (pier faces) (pier retns.) |
| 2/<br>2/2/ | 0.22<br>1.23<br>0.10<br>1.23 | Paintg. g.s., girth > 300, ① primg. ct., ② u/cs. oil paint & ① matt. fin. to pla. walls. (pier faces) (pier retns.) |
| 2/2/ | 1.23 | Rdd. L to pla. rad. 10-100. (piers) |
| 3/<br>2/2/ | 1.45<br>1.45 | (corners to rm.) (piers) |
| | | **Ent. Porch** |
| 2/ | 2.05<br>2.77 | Ct. & sd. (1:4) backg. to walls, width > 300, trowelled 10 th. on bwk., ext.<br>&<br>Renderg. to walls, width > 300, Tyrolean fin. of 'Cullamix' appld. by machine to ct. & sd. backg. (m/s), ext. |

Deductions are necessary as the voids are on the boundary of the measured area (SMM General Rules 3.4). Curved work to in situ finishings is taken separately, stating the radius measured on face (SMM M20. M5).

The paintwork is not separately classified.

In situ finishings to surfaces ≤ 300 mm wide are kept separate and measured in metres (SMM M20.1.2.1.0), but this distinction does not apply to the paintwork, as the painting is not to isolated surfaces (SMM M60.1.0.1.0). The height is adjusted for the beam intersections as the voids are on the boundary of the measured area.
Rounded angles are measured in metres, when in the radius range 10-100 mm (SMM M20. 16.0.0.0 and M20.M7. Note the variations in height resulting from the 225 × 225 mm beam.
The adjustment of finishings for the window and doors will be taken when measuring the joinery.
Backings to walls are measured in a similar manner to screeds (SMM M10.1.1.1.0), and the Tyrolean finish is measured using the same rules (SMM M20.1.1.1.0).
External work must be so described (SMM M10/20. D1).

12.6

## FINISHINGS (Contd.)

| | | | |
|---|---|---|---|
| | | Ent. Hall | The dimensions are extracted from those previously calculated for the ceiling of the Entrance Hall. |
| | | Gth. of panellg. | |
| | | 2/4·855   9·710 | |
| | |            2·755 | |
| | | retn. to scrn.  155 | |
| | |          12·620 | |
| | 12·62 | Rigid sheet ling. to walls, width > 300, 4 th. pre-finished Iroko veneered plywd. decorative panellg. in sheets 2·44 × 1·22 m w. random V jt. on face to dado, fxd. w. adhesive to swd. battens (m/s). | The wall lining to the dado is measured in m² in accordance with SMM K11.1.1.1.0 with a dimensioned description and giving details of the type of sheet, thickness, method of jointing, nature of background and method of fixing, where not at the discretion of the contractor (SMM K11. S1-13). |
| | 1·22 | | |
| | | 406)4·855     4·855 | |
| | |        12+1         2·755 | |
| | | 406)2·755     7·610 | |
| | |         7+1 | |
| 8/13/ | 1·22 | Individl. suppts. (vert. battens) 50 × 19 impregntd. sawn swd. plugged (top & bott.) to bk. walls. | Vertical battens are fixed at 406 mm centres and at the top and bottom of the panelling. The battens are measured in metres, with a dimensioned overall cross-section description as SMM G20.13.0.1.0 and the relevant particulars from SMM G20. S1-9. The term 'supports' includes battens (SMM G20. D7). |
| 2/ | 7·61 | | |
| | | 4·855 | |
| | | add retn. to scrn. 155 | |
| | | 5·010 | |
| 2/13/ | 1·22 | Ditto. fixd. to conc. (vert. battens) blk. ptn. (top & bott.) | Two additional vertical battens are dotted on to support the return length of panelling to the screen. The battens fixed to the brick walls will require plugging, while those fixed to the concrete block partition can be nailed direct, and so they need to be kept separate. The height of the wall plaster above the panelled dado is calculated in waste. |
| 2/ | 5·01 | | |
| | | ht. of pla. | |
| | | less                   2·770 | |
| | | dado  1·220 | |
| | | sktg.     225 | |
| | | rail to dado  32   1·477 | |
| | |                                1·293 | |
| | 7·61 | Pla. to walls, width > 300, in 2 cts., 13 th. w. base ct. of Thistle browning & sd. (1:2) 11 th. & finishg. ct. of fin. pla. 2 th., stl. trowelled to bwk. & Painting g.s. gth > 300, ① primg. ct., ② u/cs oil paint & ① matt fin. to pla. walls. | Plasterwork to brick and concrete block walls is kept separate, in accordance with SMM M20. S5 and General Rules 8.2. The plaster particulars contain the appropriate details listed in SMM M20. S1-8. This item is followed by the decorations, which include the relevant particulars listed in SMM M60. S1-8. |
| | 1·29 | | |

12·7

| | | | |
|---|---|---|---|
| | 5·01 | Pla. to walls, width > 300, in 2 cts. 13 th. a.b.d. to conc. blkwk. & Paintg. g.s., gth. > 300, ① primg. ct., ② u/cs oil paint & ① matt fin. to pla. walls. | The adjustments of wall finishings for the areas occupied by doors will be made when measuring the internal doors, probably working from a schedule containing all the necessary particulars. |
| | 1·29 | | |
| 3/ | 1·29 | Rdd. L. to pla. rad. 10-100. (int. Ls | Where rounded angles are in the 10-100 mm radius range, they are taken as linear items and so classified (SMM M20. 16.0.0.0 and M20. M7). Dado rails are measured in a similar manner to skirtings and picture rails in metres, with a dimensioned overall cross-section description, including the method of fixing where not at the discretion of the contractor (SMM P20.1.1.0.0 and P20. S1-9). As the rail is ≤ 0·003 m² in sectional area, ends, angles, mitres and intersections are deemed to be included (SMM P20. C1). The varnishing follows classified as SMM M60.1.0.2.0. |
| | | dado rail | |
| | 5·01 | Dado rl. 32 × 32 wrot Iroko mo., fxd. w. 13 × 32 impregnated sn. swd. spld. grds. | |
| | 7·61 | Dado rl. & grds. a.b. plugd. to bwk. | |
| | 5·01 | Clear finishg., gen. isoltd. surfs., wd., gth. > 300, 2 cts. polyurethane varnish. | |
| | 7·61 | | |
| | | cornice | Cornices are measured in metres, for the length in contact with the base, stating the girth as SMM M20.19.0.1.0, with ends, internal angles, external angles and intersections enumerated as extra over the work in which they occur (SMM M20. 23.1-4.1.0). Fibrous plaster cornices and coves are measured in accordance with SMM M31.10 & 12.0.1.0. Fibrous plaster is however seldom used nowadays and so no worked example is provided. The mean girth of the cornice is calculated in waste as it is required for painting purposes. |
| | 7·61 | Pla. cornice, gth: 125, gyp. pla. class B to BS 1191 Pt.1, to bk., blk. & conc. surfs., fltd. fin. | |
| | 4·86 | | |
| 2/ | 1 | E.O. for int. L to pla. cornice. & E.O. for end to do. | |
| | | 7·610 | |
| | | 4·855 | |
| | | 12·465 | |
| | | less corners 2/85  170 | |
| | | 12·295 | |

12·8

## FINISHINGS (Contd.)

| | | | |
|---|---|---|---|
| 12.30 | Paintg. gen. isoltd. surfs. gth ≤ 300, ① primg. ct., ② u/cs oil paint & ① matt fin. to pla. cornice. | | It is assumed that the cornice will not be continued around the return wall to the screen. The cornice is separately classified for painting purposes as it is to a base different from that of the walls and ceiling and is assumed to be of a different colour, hence it becomes an isolated surface with a girth ≤ 300 mm. |
| 12.30 0.09 | Ddt. pla. clg. ltwt. gyp. pla. in 2 cts. to conc. a.b.d. & Ddt paintg. g.s, gth. > 300, ① primg. ct., ② u/cs oil pt. & ① matt fin. to pla. clg. | | It is necessary to adjust the plaster and paintwork to the ceiling for the area occupied by the cornice. |
| 7.61 0.09 | Ddt. pla. to walls in 2 cts. 13 th. to bwk. a.b.d. | | The adjustment of the wall plaster and paintwork for the area of the cornice follows. |
| 4.86 0.09 | Ddt. pla. to walls, 2 cts. 13 th., to conc. blkwk. a.b.d. | | Regardless of the area involved, the wall and ceiling finishings will always be deductable as the cornice is on the boundary of each area and does not constitute a void (SMM General Rules 3.4). |
| 12.47 0.09 | Ddt. paintg. g.s., gth. > 300, ① primg. ct., ② u/cs oil pt. & ① matt fin. to pla. walls. | | |

### Cloaks & Telephone Cubicle

As the specification of the wall plaster and the paintwork to the cloaks and telephone cubicle are the same, the two compartments are combined for taking off purposes, to avoid the duplication of items. No adjustment is made for the height of the skirting as plastering may be continued behind the skirting.

| | |
|---|---|
| 2.28 2.77 | Pla. to walls, width > 300, gyp. pla. 13 th. in 2 cts. to BS 1191 Pt.1 of Thistle browning pla. & sd. backg. 11 th. & finishg. ct. of class B pla., 2 th., stt. trowld. to bwk. (clks. |

```
        tel. cub.        clks.
         1·100          2·280
           875          1·000
      2/ 1·975            350
         3·950          3·630
```

| | | |
|---|---|---|
| 3.63 2.77 3.95 2.77 | Pla. to walls a.b. to conc. blkwk. (clks. (tel. cub. | Work on differing bases is kept separate (SMM M20.S5). |

12·9

| | | |
|---|---|---|
| 0.65 2.77 | Ditto. to xtg. bk. walls (clks. & R.o.j. of xtg. bwk. to form key for pla. | Work on existing surfaces is so described (SMM General Rules 13.1-2). Surface treatments to masonry are measured in m², stating the type and purpose and type of wall (SMM F10. 26.1.1.0). With new work this is deemed to be included (SMM F10. C1d). |

```
              ht.        2·770
         less sktg.        150
                         2·620
```

| | | |
|---|---|---|
| 4/ 2.62 4/ 2.62 | Rdd. L to pla. rad. (clks. 10 – 100. (tel. cub. | Rounded angles to the wall plaster form separate items, where they are in the 10-100 mm radius range (SMM M20. 16.0.0.0 and M20.M7). The lengths are adjusted for the skirtings, but further adjustments will be needed when measuring the doors. The girth of the wall surfaces to the cloaks is calculated in waste from the figured dimensions on the drawing. The height is adjusted for the skirting. Decorations are measured in accordance with SMM M60.1.0. 1.0 and giving the relevant particulars from SMM M60. S1-8. The girths of the tiled wall surfaces to the gents and ladies toilets, including the cubicles, are calculated in waste. It is important to insert all preliminary calculations, no matter how trivial they may seem. In the case of the gents toilets the lengths of the wall returns are compensatory. The girths of the lobby and toilet cubicles are calculated in waste, referring back to the ceiling dimensions where appropriate. It might be considered better to precede the wall tiling by the backing, following the order of construction. |

```
                 clks. (gth.)
                       2·280
                       1·000
                  2/   3·280
                       6·560
```

| | | |
|---|---|---|
| 6.56 2.62 3.95 2.62 | Paintg. g.s., gth. > 300, seal & 2 cts. of emulsn. pt., pla. walls. (clks. (tel. cub. | |

### Toilets & Lobby
#### Gents toilet

```
                       3·310
                       3·000
                  2/   6·310
                      12·620
```

#### Ladies toilet

```
  less   4·285        2·785
  cubs.  1·500        3·000
         2·785   2/   5·785
                     11·570
```

#### Cubicles (gents toilets)

```
                       1·350
                         700
                  2/   2·050
                       4·100
```

#### Cubicles (ladies toilets)

```
                       1·450
                         713
                  2/   2·163
                       4·326
```

12·10

# FINISHINGS (Contd.)

| | | | |
|---|---|---|---|
| | | | Lobby |
| | | | 1.400 |
| | | | 875 |
| | | 2/ | 2.275 |
| | | | 4.550 |

| | | |
|---|---|---|
| | 12.62 | Ceramic tiling, to walls, plain, width > 300, (gents toilet) |
| | 2.77 | 152 x 152 x 6 th. |
| | 11.57 | cushion edged glazed (ladies toilet) |
| | 2.77 | tiles to BS1281, sym. |
| 2/ | 4.10 | layout, fxd. w. adhesive, |
| | 2.77 | & jtd. & ptd. w. white (gents cubs.) |
| 4/ | 4.33 | ct. on ct. & sd. backg. (m/s). |
| | 2.77 | & (ladies cubs.) |
| | 4.55 | Ct. & sd. (1:4) backg. (lobby |
| | 2.77 | to walls fltd. 10 th. to conc. blk. walls. |

There is no skirting to deduct when calculating the height of the wall tiling, as the tiling extends to the floor (see schedule of finishings).
The description of the ceramic tiles contains the appropriate particulars listed in SMM M40. 51-8, including the kind and quality of materials, size, shape and thickness of units, nature of base, method of fixing, and treatment and layout of joints. The cement and sand backing is measured for the full height, as a separate item, in accordance with SMM M10.1.1.1.0, and particulars as SMM M10.51-7.

| | | |
|---|---|---|
| | 1.91 | Ddt. ct. & sd. backg. (gents toilet) |
| | 2.77 | to conc. blk. walls. |
| | | & |
| | | Add ditto. to xtg. bk. walls |
| | | & |
| | | Add r.o.j. of xtg. bwk. to form key for ct. & sd. backg. |
| 2/ | 2.77 | E.O. ceramic tilg. for (gents toilet) spec. tile with 1 rdd. edge. |

Adjustments are made for the backing on different bases, as required by SMM M10.55. This is for the backing to the existing wall in the gents toilet, while the remainder is to new block walls.
An additional item is also needed to cover the preparation of the existing wall to receive the backing (SMM F10.26.1.1.0).
Special tiles are given in metres as extra over the work in which they occur (SMM M40.15.1.1.0), and the rounded edge tiles have been included in this category, despite the inclusion of paragraph M40.C2, whereby work to walls, ceilings, beams and columns is deemed to include internal and external angles and intersections ≤ 10 mm radius.
Locational notes in waste help to identify the measured items.

---

| | | |
|---|---|---|
| | | Sktgs. |
| | | Treatment Rm. |
| | | straight runs 20.600 |
| | | add pier faces 2/215 430 |
| | | pier retns. 4/102.5 410 |
| | | 21.440 |
| | | less drs. & archves. |
| | | 1.900 |
| | | 900   2.800 |
| | | 18.640 |

| | | |
|---|---|---|
| | 18.64 | Sktg., black in situ terrazzo, 10 th. & 150 hi., fin. flush w. wall terrazzo above & w. 20 rad. cove to terrazzo margin 16 th. & 110 wide at junctn. w. flr., on ct. & sd. bed & backg. (m/s). |
| 1/4/22/7/ | 3.23 | Ditto. curved to rad. of 3.23 m. |

| | | |
|---|---|---|
| | | Entrance Hall |
| | | sktgs. w. grds. plugd. to bwk. |
| | | 7.610 |
| | | less dr. & archve. 900 |
| | | 6.710 |
| | | sktgs. fxd. to blkwk. |
| | | less 5.010 |
| | | 3 drs. & archves. 2.500 |
| | | 2.510 |

| | | |
|---|---|---|
| | 2.51 | Sktg. 38 x 225 wrot Iroko, chfd., fxd. w. 13 x 32 impregnated sn. swd. spld. grds. to blkwk. |
| | 6.71 | Sktg. & grds. a.b. plugd. to bwk. |

The length of the skirting is calculated in waste, starting with the overall length of straight runs, excluding piers, extracted from the wall finishing dimensions.
The pier faces are added as these are not taken separately and the widths of doors and architraves deducted. Alternatively the latter adjustments could be taken with the doors but it is probably simpler to take them here.
Skirtings are measured in metres stating the height or width and thickness or a dimensioned description (SMM M10.13.0.1-2.1-6).

Curved work is taken separately stating the radius measured on face (SMM M10.M5). Ends and angles to in situ skirtings are deemed to be included without the need for separate enumeration (SMM M10.C10).
The length of skirting to the entrance hall is taken from the length of wall panelling previously calculated, with deductions made for the four doors and adjoining architraves. Wood skirtings are measured in metres stating the dimensioned overall cross-section description (SMM P20.1.1.0.0). The description will include the method of fixing where not at the discretion of the contractor (SMM P20.S8).
It is necessary to distinguish between skirtings with grounds plugged to brickwork and those with grounds nailed to blockwork.

## FINISHINGS (Contd.)

| | | | |
|---|---|---|---|
| | 2·51 | Clear finishg., gen. isoltd. surfs., wd., gth.≤ 300, 2 cts. of polyurethane clear varnish. | The varnishing to the skirting is measured in accordance with SMM M60.1.0.2.0. |
| | 6·71 | | |
| 3/ | 1 | E.O. sktg. for L. | Angles, ends, mitres and intersections to skirtings are enumerated if the skirting has a sectional area exceeding 0·003 m² (SMM P20.C1). |
| 10/ | 1 | E.O. sktg. for end. | |

```
              Clks. & Telephone
                   Cubicle
                    Clks.
        2/2·280    4·560
                   1·000
        less       5·560
        dr. & archves.  850
                   4·710
```

Sktg. on grds. plugd. to bwk. & fxd. to blkwk.

```
                   4·710
new bk. wall  2·280
xtg. bk. wall   650  2·930
         fxd. to blkwk.  1·780

              Tel. Cubicle
                   3·950
less dr. & archves.  800
                   3·150
```

| | | | |
|---|---|---|---|
| | 1·78 | Sktg., 25 x 150, wrot swd., fxd. w. 13 x 22 impregnated sn. swd. grds. to blkwk. (clks.)(tel. cub.) | All timber must be adequately described, including stating whether sawn or wrot (SMM P20.S1). |
| | 3·15 | | |
| | 2·93 | Sktg. & grds. a.b. plugd. to bwk. (clks.) | It is necessary to separate the skirting items according to whether the grounds require plugging (brickwork base) or not (blockwork base)(SMM P20.S8). |
| | 4·71 | Paintg. gen. isoltd. surfs., wd., gth. ≤ 300, k.p.s., ② u/cs & ① hd. gloss pt. (clks.)(tel. cub.) | The description of paintwork is to state the number of undercoats and finishing coats, and the surface finish (SMM M60.55-6). |
| | 3·15 | | |

12 · 13

## Floors
### Treatment Rm.

```
                     3·550
       add beam       225
                     3·775

                     3·225
       add beam       300
                     3·525
```

Screed to flr., lev., 34 ct. & sd. (1:3) trowld., on conc. base.

Calculate dimensions of rectangular areas of floor finishings to Treatment Room in waste, using as a starting point the dimensions already calculated for the ceiling.

| | | |
|---|---|---|
| | 6·75 | |
| | 3·78 | |
| | 3·53 | |
| | 3·23 | |
| ¼/22/7/ | 3·23 | |
| | 3·23 | |
| | 1·80 | (swing drs. opg. |
| | 0·10 | |
| 2/ | 0·22 | Ddt. last (piers |
| | 0·10 | |

Screeds are measured in m² as SMM M10.5.1.1.0, giving the appropriate particulars listed in SMM M10.S1-7 (no width range required). The actual type of finish to be received is included in the description. Door openings are normally included at this stage.
The work to the curved perimeter does not require special mention.
To determine the area of terrazzo paving it is necessary to deduct the projection of the base of the skirting from the overall dimensions as used for the measurement of the screed.

```
                     6·750
less sktgs. 2/100    200
                     6·550
                     3·775
less sktg.           100
                     3·675
                     3·525
less sktg.           100
                     3·425
                     3·225
less sktg.           100
                     3·125
```

In situ terrazzo flr., lev., 16 th. black & white, ct. & marble chippgs.(1:2), ld. alternately in panels abt. 500 x 500 between ebonite dividg. strips (m/s), trowelling to smth. fin. & pol. on & inc. bldg. paper underlay, ld. on trowelled screed (m/s).

| | | |
|---|---|---|
| | 6·55 | |
| | 3·68 | |
| | 3·43 | |
| | 3·13 | |
| ¼/22/7/ | 3·13 | |
| | 3·13 | |
| | 1·80 | (swg. drs. opg. & sktg. |
| | 0·10 | |
| 2/ | 0·22 | Ddt. last. (piers |
| | 0·10 | |

A full description of the terrazzo paving is required giving the appropriate particulars listed in SMM M10.5.1.1.1-2.
The building paper underlay is included in this item but not the ebonite dividing strips.

A deduction has to be made for the area occupied by the piers as they are on the boundary of the paved area (SMM General Rules 3.4).

12 · 14

| | | | | | | | |
|---|---|---|---|---|---|---|---|
| FINISHINGS | | (Contd.) | | | | | |

```
                          dividg. strips
                             7·000
              less sktgs. 2/100    200
                             6·800
                     3·550   6·740
              less sktg.  100    5·350
                     3·450  2)12·090
              add pier   215    6·045
                     3·665
                             6·750
              less sktgs. 2/100    200
                             6·550
                     pier   6·660
                     3·125   5·350
              add beam   300  2)12·010
                     3·425    6·005
```

| | | | |
|---|---|---|---|
| 9/ | 6·80 | | Dividing strip, 6 x 16 ebonite, bedded in c.m. (1:3) between terrazzo flr. panels. |
| 5/ | 6·05 | | |
| 9/ | 6·55 | | |
| 5/ | 6·01 | | |
| | 3·67 | | |
| | 3·43 | | |
| 2/ | 0·10 | | (pier retns. |
| 4/ | 0·10 | | (swg. drs. opg. |
| ¼/22/7/ | 3·13 | | Dividg. strip a.b. curved to rad. of 3·13 m. |

### Ent. Porch

| | | |
|---|---|---|
| 2·60 | | Screed to flr., lev., 28 ct. & sd. (1:3), trowld., on conc. base, ext. |
| 2·05 | | |

&

Quarry tile flr., lev., plain, 152 x 152 x 22 th. brown tiles to BS 1286, Type A, bedded, jtd. & ptd. in c.m. (1:3) on trowld. screed (m/s), ext.

12·15

The lengths of the dividing strips are calculated in waste, counting the numbers of full lengths in each direction from the drawing and adjusting for the 100 mm wide margin. The lengths intercepted by the curved wall are averaged in length and the shorter lengths on two sides of the room linking up with the curved section are then taken. Dividing strips are measured in metres with a dimensioned description as SMM M10.24.7.1.0. In practice the separate measurement of the dividing strips in this situation might be regarded as superfluous and the surveyor might be tempted to include them in the description of the terrazzo paving, but this would pose problems for the estimator as he needs to know the length of dividing strip required. Additional information has been included in the description as permitted by SMM General Rules 1.1. Further additions are inserted for a pair of pier returns and the cross joints at the opening for the swing doors and line of skirting, and to the single door. Curved work is kept separate, stating the radius (SMM M10.M5). The measurements of the screed to receive quarry tiles are identical to those of the ceiling determined previously. The quarry tiles form a separate item to the screed and incorporate in the description the appropriate particulars listed in SMM M40.S1-8. There must be adequate information for the estimator to price the work.

### Ent. Hall

| | | |
|---|---|---|
| 4·86 | | Screed to flr., lev., 25 ct. & sd. (1:3) fltd. on conc. base. |
| 2·76 | | |

&

| | | | |
|---|---|---|---|
| | 1·55 | | Wd. blk. flr. lev., (ent. drs. opg. 300 x 75 x 25 th. |
| | 0·10 | | |
| 3/ | 0·75 | | Iroko wrot t.&g. (dr. opgs. to treatmt. rm., blks., herringbone, clks. & lobby pattn. w. 2 blk. |
| | 0·10 | | |
| | 0·70 | | plain margins a/rd. bedded & jtd. in (dr. opg. to mastic on fltd. (tel. cub. scrd. (m/s), w. machine sanded fin., & seal, body in & 2 cts. of wax pol. |
| | 0·10 | | |

| | | | |
|---|---|---|---|
| | 1·55 | | Dividg. strip, 6 x 25, ebonite, bedded in c.m. (1:3). |
| 3/ | 0·75 | | |
| | 0·70 | | (dr. opgs. |

### Clks.

| | | |
|---|---|---|
| 2·28 | | Screed to flr., lev., 40 ct. & sd. (1:3) stl. trowelled, on conc. base. |
| 1·00 | | |
| 0·75 | | |
| 0·10 | | & (dr. opg. to gents toilet |

Plastics flr., width > 300, lev., 300 x 300 x 2 th. PVC tiles to BS 3261 Type A fxd. w. adhesive on trowelled screed (m/s).

| | |
|---|---|
| 0·75 | Dividg. strip 6 x 16, ebonite, bedded in c.m. (1:3). (dr. opg. |

All work is deemed internal unless described as external (SMM M40.D1), hence the addition of 'external' to the description.

Screeds include the appropriate particulars listed in SMM M10.S1-7. The description shall also include the method of surface treatment, such as floated or trowelled and the nature of the base.

The description of wood blocks incorporates the appropriate particulars listed in SMM M40.S1-8, including the layout of joints. All cutting is deemed to be included (SMM M40.C1d). The description is to include all preparatory work and the nature of the finished surface, including any sealing/polishing.

Linear item as SMM M40.16.4.1.0, with a dimensioned description and giving the method of fixing (SMM M40.S9). The minimum depth is taken as the thickest adjoining floor finish.

The thickness of the bed is varied according to the thickness of the finishing to provide a level surface throughout. The overall thickness is 50 mm. The PVC floor tiling is measured in accordance with SMM M50.5.1.1.0, including the prescribed width range.

The flooring to the cloaks is followed by the dividing strip in the door opening to the gents toilet. Alternatively all strips in door openings could be taken together at the end of the floor finishings.

12·16

| | | | | | | |
|---|---|---|---|---|---|---|

FINISHINGS (Contd.)

<u>Tel. Cubicle</u>

| | |
|---|---|
| 1·00 | |
| 0·88 | |

Screed to flr., lev., 47 ct. & sd. (1:3) stl. trowelled on conc. base.

&

Cork flrg., width > 300, lev., 300 x 300 x 3·2 th. cork tiles, Cork-o-Plast, checker tiles nr. 13, semi-bright fin., butt jtd. & fxd. w. copad adhesive on trowelled scrd. (m/s).

The trowelled screed and the cork tile finishing to the telephone cubicle form two separate items, with the cork tiles fully described in accordance with SMM M50.5.1.1.0, including the prescribed width range classification.

<u>Ladies & Gents Toilets & Lobby</u>

| | |
|---|---|
| 3·31 | |
| 2·55 | |
| 1·91 | |
| 0·45 | |
| 4·29 | |
| 3·00 | |
| 1·45 | |
| 0·88 | |

Screed to flr., lev., 40 ct. & sd. (1:3), trowld., on conc. base.    (gents toilet

&    (ditto. recess

Ceramic tiled flr., lev., plain, 152 x 152 (ladies x 9·5 th. fully     toilet vitrified ceramic tiles to BS 1286, w. (lobby 3 wide jts. bedded, jtd. & ptd. w. c.m. (1:3) on trowld. scrd. (m/s).

The flooring to the gents and ladies toilets and the lobby can be taken together as they all have the same finishing. The boundary of a floor finish is taken to coincide with a door so that the joint is not visible when the door is closed. The length of the lobby has been increased to include the finish to the door opening into the ladies toilet. No further dividing strips are required.

12·17

# 8 MECHANICAL SERVICES

## GENERAL BACKGROUND

The measurement of mechanical services installations requires a detailed knowledge of technology. The drawings from which the quantity surveyor must work are those that are prepared by the consulting services engineer. Consulting engineers are required by their scale of charges to prepare drawings and specifications 'sufficient to obtain tenders'. The drawings are schematic only, as it is trade practice for the contractor to prepare all necessary working drawings and to include in his tender for a complete working installation. The exact routeing of pipework and ductwork is often left to the craft operatives doing the work and the contractor will have in mind such factors as the location of other services, restrictions on space and the ease of maintenance of the completed work. In essence, the quantity surveyor must put himself in the position of the operative and include in the bill of quantities all items necessary for the complete installation. A good practical knowledge of the technology of services installations is therefore essential.

Many quantity surveying practices employ engineers within their organisations either to give advice to the taker off or to take off the quantities themselves. It is, however, generally believed that the traditionally trained quantity surveyor can acquire the necessary additional knowledge of technology by private study, attending courses, observing site installations and taking measurements on site and is then competent to prepare accurate bills of quantities for mechanical services.

## MEASUREMENT PROCEDURES

The first task in taking off, as with any other work section, is fully to study and understand the drawings and specifications provided. A study of these documents will inevitably lead to queries. These queries must first be scrutinised by the in-house engineer and/or the appropriate partner to eliminate any obvious discrepancies and must then be submitted to the consulting services engineer in the usual way. The query/answer procedure is essential for the preparation of good bills. The consulting services engineer may not be very familiar with the process and may regard the completion of the query sheets as an additional burden for which he receives no payment. Skill, tact and diplomacy are needed to limit the number and frequency of queries and to stress their benefits to the engineer, namely, that they help to remove anomalies that could cause future problems on site.

The taker off will need to make sensible approximations on occasions, where the method of working is left to the operative on site. Two examples of such approximations are firstly whether to measure made bends or fittings and secondly what allowance to make for co-ordination of services. In the first instance one approach in practice is to measure bends only in the bill of quantities and to insert a suitable preamble to the effect that the contractor must allow in his prices for the provision of made bends or fittings as required. Another option is to state in the preambles the proportion of made bends to fittings for different classes of pipework and that the contractor must allow in his prices for any additional fittings required. The consulting services engineer should be consulted in the preparation of this schedule, and/or be requested to give guidance on likely requirements.

With regard to co-ordination it is difficult to give specific guidance. Where co-ordinated drawings are provided by the engineer — the exception rather than the rule — many of the problems will have been resolved and a minimal inclusion of additional pipework and bends/fittings will be sufficient. Where no co-ordinated drawings are available, however, and where services run in congested areas, the problems of co-ordination may be such that services may need to be substantially diverted requiring additional pipework and possibly additional builder's work. Each scheme must be judged on its merits and, in consultation with the engineer, suitable provision should be made. Co-ordination items are best included in the bills either as approximate quantities or provisional sums in accordance with SMM General Rules 10.1 and 3.

## APPROACH TO MEASUREMENT

Having acquired adequate information and a full understanding of the scheme, taking off can begin. Firstly, the drawings should be coloured up using a suitable colour code to illustrate the various services to be measured. The drawings are mainly in the form of floor plans and it is often helpful to draw a sketch of complicated sections of the work to be taken off in isometric projection to illustrate the full extent of the work involved. These sketches provide a valuable record of the work measured and are often useful for final account purposes.

A good workmanlike method of booking dimensions and striking through work on the drawings is essential to make sure that nothing is missed. Since the work is of a repetitive nature, a schedule approach is often favoured with the items listed across the top of a sheet of abstract-sized paper and with location information given on the left-hand side. The use of schedules however has the disadvantage that the dimensions are less easy to follow and read. The traditional quantity surveying approach tends to produce reams of paper and to be time-consuming. A sensible compromise used by some practices is to take off on traditional dimensions paper with abbreviated descriptions and to use a cut and shuffle abstract system which will cope effectively with the constant repetition of items.

A good library of reference information is absolutely essential. This will include textbooks on technology, trade literature, relevant British Standards and Codes of Practice, the current edition of the IHVE (Chartered Institute of Building Services) guide and the current edition of the Heating and Ventilating Contractors' Association's Specification for Sheet Metal Ductwork.

The process of measurement is relatively straightforward, comprising enumerated items of plant at the source, a connecting network of pipework or ducting measured linearly, with enumerated fittings taken as extra over pipes and ducting, enumerated ancillaries, and finally enumerated items for the emission plant and equipment. Special supports and sleeves for pipes and ducting each generate further enumerated items. Assuming that the installation to be measured is fully understood and sufficient specification information is available, no great difficulty should be experienced provided a methodical approach is adopted throughout.

Work in plant rooms is identified separately because of the restricted working conditions (SMM Y10/20/30.M2). Everything necessary for jointing is deemed to be included (SMM Y10/20/30.C1), and full requirements of materials shall be given as prescribed in SMM Y10/20/30.S1–6. Thermal insulation to pipelines and ducting is measured in metres giving the nominal size of the pipeline or ducting, while insulation to equipment is enumerated giving the overall size or measured in m$^2$ (SMM Y50.1.1, 3 & 4.1.0).

Section Y of SMM7 must be read in conjunction with the appropriate sections of the *Code of Procedure*. The *Code of Procedure* gives examples of items which are not included in the text of SMM7. For instance, Y10/11:2.4 in the Code states that examples of pipe fittings would include bends, springs, offsets, swan necks, Y-junctions, double Y-junctions, blank flanges, puddle flanges, bushes, reducers, elbows, twin elbows, tees, crosses and unions. The special significance of this item is that pipe fittings ≤ 65 mm diameter are grouped together irrespective of type, stating the number of ends. With larger fittings the type is stated (SMM Y10.2.3–4.2–6.1–2). In section Y10:8.1, the Code gives examples of pipework ancillaries which include draw-off taps, stop valves, control valves, regulating valves, safety valves, reducing valves, non-return valves, drain cocks, stop cocks, air cocks, mixing valves, steam traps, strainers, gauges and thermometers, and automatic controls.

The worked example in this chapter provides a fully annotated take off covering the measurement of services for a particular project, selected to give a good range of different features. Care must be taken not to apply the descriptions and other data to another project, without careful reference to the particular specification and engineering requirements of the scheme in hand.

# Mechanical Services

## OTHER MEASUREMENT ASPECTS

### Work sections

The rules for the measurement of mechanical services given in Work Group Y are billed under separate work sections as listed in Appendix B of SMM7. The requisite work sections are as follows:

R Disposal systems, which include sewage pumping and refuse chutes.

S Piped supply systems, which include cold water, hot water, steam, fire hose reels, dry risers, wet risers and sprinklers.

T Mechanical heating/cooling/refrigeration systems, which include gas/oil fired boilers, coal fired boilers, heat pumps, solar collectors, low temperature hot water heating, steam heating, warm air heating and central refrigeration plant.

U Ventilation/air conditioning systems, which include toilet and kitchen extracts, smoke extract/smoke control and various forms of air conditioning.

### Pipework generally

Pipes are classified under appropriate headings, such as hot water supply (S11 in Appendix B of SMM7 and the *Common Arrangement*), and measured over all fittings and branches in metres, stating the type, nominal size, method of jointing and type, spacing and method of fixing supports, and distinguishing between straight and curved pipes (SMM Y10.1.1.1.0 and Y10.M3). Pipes are deemed to include joints in their running length (SMM Y10.C3), and the provision of everything necessary for jointing (SMM Y10.C1), without the need for specific mention. The type of background to which the pipe supports are fixed will be classified in the categories listed in SMM General Rules 8.3.

Details of the kind and quality of materials used in the pipes, gauge and other relevant particulars listed in SMM Y10.S1–6, are likely to be included in preamble clauses or a project specification.

Made bends, special joints and connections, and fittings such as Y-junctions, reducers, elbows, tees and crosses, are all enumerated as items extra over the pipes in which they occur (SMM Y10.2.1–4). In the case of special joints, the type and method of jointing is to be stated and they comprise joints which differ from those generally occurring in the running length or are connections to pipes of a different profile or material, connections to existing pipes or to equipment, appliances or ends of flue pipes (SMM Y10.D2).

Pipe fittings ≤ 65 mm diameter are classified according to the number of ends, while those of larger diameter are described. The method of jointing is stated where different from the pipe in which the fitting occurs.

Valves and cocks are classified as pipework ancillaries and are enumerated, stating the type, nominal size, method of jointing, type, number and method of fixing supports and type of pipe to be connected (SMM Y11.8.1.1.0). Those located in ducts or trenches are each kept separate and so described.

Cutting holes through the structure for pipes and making good surfaces are enumerated, stating the nature and thickness of the structure and the shape of the hole, and classifying the pipes as to size in accordance with SMM P31.20.2.1–3.2 & 4; for example pipes ≤ 55 mm nominal size, 55–110 mm and > 110 mm. The cutting of holes for pipes is best picked up when the various lengths of pipework are being taken off, rather than leaving all the holes to be taken off after the pipework has been measured complete. By contrast, painting of pipes may often, with advantage, be left to the end of the taking off.

### Adequacy of measurement

It is quite usual for parallel flow and return pipes to be shown by a single line on the engineer's schematic drawings and annotated F and R with a note of the dissimilar sizes that often occur. When measuring, adequate allowance must be made for bends, circumventing obstructions, and even for cold feed vent and air release pipes, plant room drains and pump by-passes, where not shown in detail.

### Equipment

When measuring mechanical equipment, such details as type, size and pattern, rated duty, capacity, loading as appropriate and method of fixing are stated. Specification cross references are often inserted for mechanical equipment as provided for in SMM Y20/40.1.1.1.0. However, the excessive use of cross references to the specification can be inconvenient to the estimator and fuller descriptions in the bill may sometimes form the better approach, although this runs contrary to the wider use of project specifications. Where insufficient data are available, the work can be covered by PC or provisional sums or a bill of approximate quantities may be prepared.

Examples of equipment are listed in the *Code of Procedure* and include boilers, generators, water treatment and pressurisation plant, tanks, cylinders, calorifiers, pumps, compressors, fans, filters, humidifiers and refrigeration units.

### Air ductlines

Ducting is classified as to whether straight, curved stating radii or flexible, and giving the type, shape, size, method of jointing and spacing and method of fixing supports, and background, and is measured in metres as SMM Y30.1.1.–5.1.1. Like pipes, it is measured over all fittings and branches (SMM Y30.M3), and is deemed to include joints in running lengths and stiffeners (SMM Y30.C3).

Items measured extra over the ducting in which they occur include the following:

(1) lining ducting internally in metres, stating the type and thickness of lining material and internal size of ducting (SMM Y30.2.1.1.0);

(2) special joints and connections, as described in SMM Y30.D2, enumerated, stating the type, size, ducting size and method of jointing (SMM Y30.2.2.1.1);

(3) fittings, such as stop ends, bends, offsets, diminishing pieces, change of section pieces and junction pieces; access openings and covers or doors; nozzle outlets; and test holes and covers are each enumerated, stating the type as SMM Y30.3–6.1.1.

Ancillaries to ducting, such as grilles, diffusers, dampers, shutters, cowls, terminals, roof ventilators, attenuators and anti-vermin screens, are enumerated giving the information prescribed in SMM Y30.4.1.1.0, while breaking into existing ducting is given as an item, stating the type, size and location of duct and purpose of breaking in (SMM Y30.5.1.1.1–4).

Ducting sleeves are enumerated and classified and described as for pipes (SMM Y30.7.1–2.1.1–2).

### Pipe and ducting supports

Pipe and ducting supports which differ from those given with pipe or ductlines are separately enumerated, giving details of the nominal size of pipe or shape and size of duct, type and size of support, method of fixing pipe or duct support and nature of background (SMM Y10.9.0.1.3 and Y30.6.0.1.3).

### Builder's work in connection with mechanical installation

Builder's work in connection with a mechanical installation is identified under an appropriate heading (SMM P31.M2). Unless identified in SMM work sections P30 and P31, all other items of builder's work associated with the mechanical installation are given in accordance with the appropriate work sections (SMM P31.M1). Where a hot water and heating installation is to be carried out by a nominated sub-contractor, items will be provided to cover any specific items of special attendance required in accordance with SMM A51.1.3.1–8.1–2, classified as either fixed or time related charges. General attendance on nominated sub-contractors is measured in accordance with SMM A42.1.16.

## WORKED EXAMPLE

The worked example covers the measurement of low pressure hot water heating and ventilation systems as illustrated on drawing 13 and described in the following extract from the specification of the engineering works, which the student is advised to study carefully.

The drawing shows more work than is actually measured in this example, but it is considered that this provides a more realistic approach and will give the student practice in identifying specific parts of the work. In like manner extracts from fan convector and grille schedules have been inserted to illustrate their usual format even though only one item is actually taken from each, to avoid considerable repetition of similar items.

## EXTRACT FROM THE SPECIFICATION OF ENGINEERING WORKS FOR THE TREATMENT BLOCK

### 1 Drawings and Documents

The Contractor shall be responsible for the preparation and supply of all detail drawings for builder's work, wiring diagrams and drawings of work done by other trades, required for the purpose of the installation and the cost of these must be included in the tender.

Within ten days of certified practical completion of the contract, the Contractor shall supply to the Engineer, two complete sets of 'as installed' drawings on heavy quality lined or tracing film and one set of half-plate photograph negatives indicating the exact position of all plant, equipment and pipe runs as actually installed.

## 2 The Building

The construction generally is of loadbearing hollow external walls with concrete ring beam and part loadbearing blockwork internal walls carried on concrete 'trench fill' foundations with a reinforced concrete ground slab.

The first floor is of 200 thick prestressed concrete floor units with a structural topping.

The roofs generally are flat and covered with asphalt on prestressed concrete units. There are small areas of bituminous felt on 'Purldeck' and timber joists.

The floor to floor heights are – ground floor 3.5 m and first floor 3.2 m. Ceiling heights are 2.8 m high throughout.

## 3 Pipework Generally

Although the runs of pipes are shown as accurately as possible on the plans, the Contractor shall be responsible for arranging the runs, both vertical and horizontal, with a view to avoiding unsightliness and ensuring ease of maintenance.

Where such arrangement involves special fittings or setting of pipes, even though not specifically shown on the plans, such fittings or sets shall be provided under this contract.

Pipes shall be fixed to give a minimum clearance between the outside of either bare pipes or lagging as follows

25 mm between pipe flanges or unions and walls.
100 mm between pipes and finished floor or ceiling.
150 mm clear of cables or conduit.
50 mm clear between pipework and ductwork.
75 mm between pipes.

## 4 Pipe Supports

All pipework is to be securely supported at intervals, not exceeding those laid down in the IHVE (CIBS) Guide Part B Table B16.3.

Additional brackets shall be provided as necessary at the beginning and ends of runs, and at junctions and bends, along with all necessary brackets and supports below heavy items of plant, such as valves, to ensure that no strain is transmitted to the pipework.

All brackets and parts of bracket assemblies shall be delivered to site with a protective coat of paint or similar. Where proprietary items are shown, these shall have the manufacturer's standard finish. All other brackets shall be given one coat of red oxide paint before delivery to site. Supports fixed externally to the building shall be galvanised.

Where pipework is exposed to view in rooms the Contractor shall supply and fix clips of the school-board pattern where pipes are adjacent to walls.

Where pipes are suspended in the ceiling voids the Contractor shall provide a mild steel channel section fixed to the structure and split ring and drop rod hangers suspended from the channel section.

Where pipes are run in floor trenches (ducts) the Contractor shall provide and the Main Contractor cast in, suitably sized mild steel angles from which the pipes shall be supported by means of drop rods and split rings or purpose-made hangers as detailed previously.

## 5 Pipe Sleeves and Floor Plates

Where pipes pass through walls, floors or ceilings, rigid PVC sleeves shall be provided to finish flush with the surface of the building fabric. The sleeves shall be of a size to accommodate the pipe so that no pipe shall touch the sleeve or the building fabric. All sleeves shall be fitted with polished chrome zinc alloy wall or ceiling plates with set screws, as manufactured by Crane Ltd or Ideal Standard Ltd. Adequate spacing shall be allowed between pipes for easy fixing of these plates which shall be cut as necessary to fit walls, corners and the like.

## 6 Draining and Air Venting (Water Services)

Drain valves shall be provided where shown on the drawings, at all points where required for adequate drainage of the apparatus and pipework, and on the 'dead' side of all isolating valves.

They shall be Hattersley Fig. 371 15 mm draining taps for pipeline drainage and Hattersley Fig. 81HU, sizes as detailed, for plant drainage. Six loose keys shall be handed to the Employer's Maintenance Engineer on completion.

## 7 Fan Convectors and Makeup Air Units

The Contractor shall supply and install in the positions indicated on the drawings fan convectors and warm air makeup units as manufactured by Dunham Bush Ltd.

The model, type and size of each fan convector is shown on the drawings.

All casings shall be constructed from heavy gauge sheet steel stiffened to prevent distortion, drumming and vibration.

## 8 Pipework and Fittings

The installation of Low Temperature Hot Water Heating shall be carried out in heavy quality mild steel tube to BS 1387:1985; black unless stated otherwise in the Specification.

All pipework in Plant Room, trenches, floor ducts and ceiling voids shall be of welded construction.

All exposed pipework shall have screwed and socketed joints in accordance with BS 21:1985 and with full length taper threads pulled up tightly, except in the Plant Room where welded pipework shall be used.

The Contractor is to allow for wire brushing all exposed pipework immediately before the painting Contractor starts work in that area.

All branches (except for air vents or drain valves) shall be of easy sweep tee type and shall, where the sizes are available, be the seamless steel butt welding branch bend type fittings. Where the branch off the main is outside the range of fittings available, the branch shall be formed from tube, carefully formed to an easy sweep.

Wherever possible, bends up to 45° shall be fire made, or cold pulled, but for larger angles seamless steel butt welded bends shall be used.

Square branches shall be used for all drain valves or air venting points.

On screwed pipework, short sweep patterns, black beaded malleable iron fittings shall be used of G. F. brand manufactured by Le Bas Tube Co. Ltd, or Crane Ltd to BS 143:1986.

### 9 Valves

| Service | Nominal Pipe Size | Manufacturer | Type |
| --- | --- | --- | --- |
| (i) Heating Mains and Plant | Up to and including 50 mm | Hattersley | Fig. 33X Wheel or lockshield screwed BSP |
| (ii) Radiators and where concealed in fan convectors | 15 mm to 25 mm | Hattersley | 'Delflo' Fig. 2407 straight pattern wheel or lockshield |
| (iii) 3-way diverting valve and motor | Up to and including 50 mm | Staefa | Fig. M3P 25 G |

### 10 Inlet Louvres

The Contractor shall supply and install input louvres to the makeup air units as manufactured by Neta-Line Air Distribution Products Ltd.

Louvres shall be type NA — 38 mm flanged framed fixed louvres with insect guard. Louvres shall be constructed from aluminium extrusions having a standard mill finish.

### 11 Ductwork

Sheet metal ductwork shall be manufactured and installed in accordance with the HVCA Publications DW/121, DW/132 and DW/112.

### 12 Ductwork and Plant Supports

Rod hangers shall consist of 10 mm diameter drop rods for ducts up to 1.2 m wide and 12 mm diameter for ducts over 1.3 m wide. The drop rods shall be secured to the roof structure via mild steel channels and expanding bolt fixings wherever possible. The brackets supported by the rods shall for rectangular ducts be fabricated from mild steel flat of the same size as the angle flanges of the duct supported. All drop rods shall be secured to the brackets with a nut and washer above and a washer, nut and locknut below.

The bracket supported by a drop rod for circular ducts shall be of the split clip type manufactured from 25 mm × 5 mm mild steel strip.

### 13 Volume Control Dampers

Rectangular volume control dampers shall be Model DDB as manufactured by Barber and Colman Ltd.

Damper frames shall be flanged and drilled suitable for direct bolting to the ductwork systems.

All dampers shall be fitted with an external operating lever and quadrant which shall be lockable in any position between fully open and fully closed.

### 14 Fire Dampers

The Contractor shall supply and install where indicated on the drawings fire dampers as manufactured by Actionaire Equipment Ltd.

Fire dampers shall be Series 201 and a fire resistant sealer shall be inserted between the duct and the damper spigot as recommended by the manufacturers.

The Contractor shall supply and fix access doors on each side of the fire dampers.

### 15 Grilles

The Contractor shall supply and install, as indicated on the drawings, grilles as manufactured by Neta-Line Air Distribution Products Ltd.

All grilles shall be standard finish and be suitable for concealed fixing.

### 16 Insulation

| Location | Service | Pipe Size | Thickness | Material | Finish |
| --- | --- | --- | --- | --- | --- |
| Where concealed from view, that is, ceiling voids and service ducts | Heating and hot water service pipework | Up to and including 65 mm | 19 mm | Long fibred rockwool preformed section and fittings secured with lacing wire | Canvas covered with three metal bands per section |

# FOR READER'S NOTES

**DRG. 13**

## LOW PRESSURE HOT WATER HEATING AND VENTILATION SYSTEMS TO GROUND FLOOR OF PART OF TREATMENT BLOCK

### GRILLE SCHEDULE

| Ref. nr. | Size | Type |
|---|---|---|
| G1 | 400 x 400 | A2/H-R |
| G2 | 400 x 400 | ditto |
| G3 | 550 x 300 | N/A/IS |
| G4 | 350 x 300 | A2/H-R |
| G5 | 600 x 400 | ditto |
| G6 | 150 x 150 | ditto |
| G7 | 250 x 250 | ditto |
| G8 | 250 x 250 | ditto |
| G9 | 150 x 150 | ditto |
| G10 | 250 x 250 | ditto |
| G11 | 1800 x 100 | ES15/Border 3/HM |

### FAN CONVECTOR SCHEDULE

| Ref. nr. | Type | Control | Additional Equipment |
|---|---|---|---|
| FC1 | J35/207 | Medium speed only | |
| FC2 | ditto | ditto | |
| FC3 | ditto | ditto | |
| FC4 | J42/226 | Medium speed. Remote stat. by others | Dunham Bush grille |
| FC5 | J20/204 | Inbuilt on/off stat. Inbuilt low water temp. stat. | |

NOTES:
1. All fan convectors on 100 high plinths by joiner
2. All transfer grilles at L.L. in doors
3. These drawings are for tender purposes only
4. The contractor shall prepare working drawings co-ordinated with other trades works
5. These drawings shall be read in conjunction with the specification of engineering works

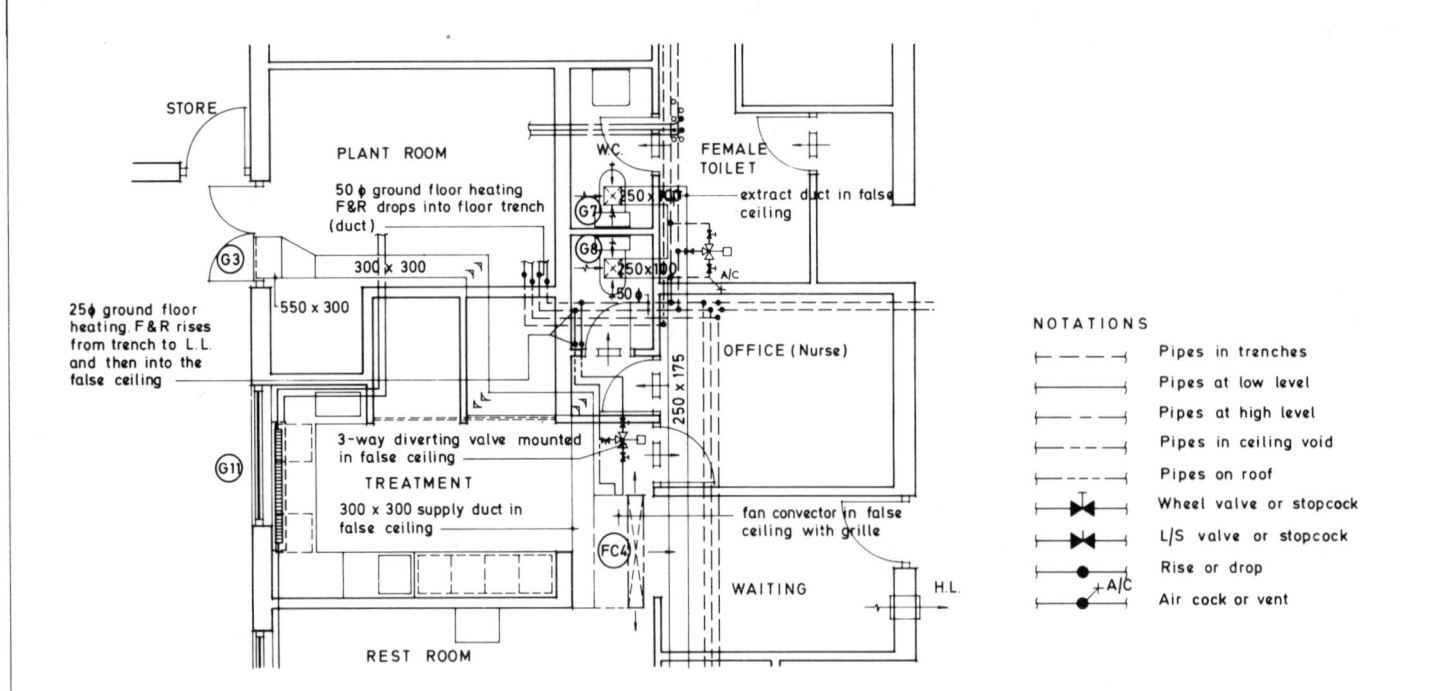

**NOTATIONS**

- Pipes in trenches
- Pipes at low level
- Pipes at high level
- Pipes in ceiling void
- Pipes on roof
- Wheel valve or stopcock
- L/S valve or stopcock
- Rise or drop
- Air cock or vent

**PLAN**   Scale 1:100

## MECHANICAL SERVICES TO TREATMENT ROOM

Note: This example includes the low temperature hot water heating flow & return branch pipes from the mains in the trench (duct) to FC4 & all the associated ductwork, as shown on Drawing 13 and described in the specification notes.

### Description of the Installations

The wks. comprise the supply, installatn. testg., regulatg., commissiong. & settg. to wk. of the follg. mech. services to the Treatment Block.

A low temperature hot water heating installatn. shall serve the bldg. operating at a constant temp. to feed the fan convectors which are provided w. automatic controls.

*A general description of the installation may be provided where it is not evident from the location drawings provided under SMM Y10/20/30.P1a, showing the scope and location of the work, including the extent of work in plant rooms. The location drawings would not normally show the mechanical services in detail. Reference to the project specification can avoid lengthy descriptions.*

A supply duct system shall be provided serving the fan convector in the Waiting Area.

*Only the section of ductwork to be measured has been described.*

### Regulations, Rules & Byelaws

The installatn. shall comply with:

a) The regulatns. & byelaws of any Statutory Body, inc. L.A.s, properly exercisg. jurisdictn. over the wks.

b) The recommendatns. as published in the current edition of the IHVE (CIBS) Guide.

*Relevant statutory and technical requirements have been inserted to assist the estimator in pricing this specialist work, although this is not required by SMM7.*

13.1

---

## EXAMPLE 13

### LOW TEMPERATURE HOT WATER HEATING INSTALLATION

Flow & retn. pipewk. from tr. to FC4
(shown in Bill Diag. Nr.1)

[Isometric sketch showing: supports, ceiling line, motorized valve, 25ø f&r, floor line, trench profile (duct), 20ø valves to FC4 with reducers, 50ø mains]

ISOMETRIC SKETCH  NTS

BILL DIAGRAM Nr 1

*Classification of work as SMM T13 in Appendix B; hence described as low temperature instead of low pressure.*

*Work in plant rooms is identified separately (SMM Y10.M2).*

*The sketch (Bill Diagram Nr 1) illustrates the actual route of the pipework to be measured. This could be included as a Bill Diagram although it is not standard practice, and non-dimensioned bill diagrams are not listed as drawn information under SMM General Rules 5, but their use could be justified under SMM General Rules 1.1, giving more detailed information than is required by the rules where necessary to define the precise nature and extent of the required work.*

*Clause 3 in the Specification is typical and states that where special fittings or sets of pipes are required additional to those shown on the drawings, they shall be provided under the contract.*

*The pipework on the drawing is shown side by side for clarity, whereas it will be fixed with one pipe above the other. It will also be noted that there are 12 bends shown on the drawing although 14 will be required for the installation.*

*Extracts from the specification are collated to draft the preambles, using the criterion that if a specification clause affects the price it shall be included in the preambles or the measured items. A typical item is clause 4 covering the protective preparation by the manufacturer of proprietary items, before delivery to site.*

13.2

## MECHANICAL SERVICES (Contd.)

|   |       |                                                                 |
|---|-------|-----------------------------------------------------------------|
|   |       | G.F. to F.F.     3·500                                          |
|   |       | flr. thickness    200                                           |
|   |       | flr. to clg.     3·300                                          |
|   |       |                   500                                           |
|   |       | len. to wall     3·800                                          |
|   |       |                                                                 |
|   |       |                  1·100                                          |
|   |       |                   400                                           |
|   |       |                  1·300                                          |
|   |       |                   300                                           |
|   |       | len. in clg.     3·100                                          |

<u>Heavy quality black m.s. tube to BS 1387 w. scrd. & socketed jts. in accordance w. BS 21, fxd. w. school board patt. pipe clips & wire brushg. exposed pipes prior to paintg. by others.</u>

| 2/ | 3·80 | Pipes, st., nom. size: 25, fxd. w. pipe clips @ 2·4 m ccs. to masonry. (rise from tr. to clg. F & R)  &  Paintg. servs., isoltd. met. surfs., gth. ≤ 300, degrease, 1 ct. calcium plumbate primer, ① u/c & ② cts. gloss. |
|---|---|---|
| 2/ | 1 | E.O. m.s. pipe, nom. size: 50, for fittg., 3 nr. ends. (reducg tee in flr. tr.) |

Pipes are measured over all fittings and branches (SMM Y10.M3).

To avoid the repetition of information a heading giving the basic specification information is essential. The specification states that where pipework is exposed it shall have screwed joints and where concealed, welded joints (clause 8). Joints in the running length are deemed to be included in the pipes (SMM Y10.C3). Standard pipe supports, and their spacing and method of fixing is given in the description of the pipework (SMM Y10.1.1.1.1.). The wirebrushing is a requirement of the specification, where pipework is exposed. The normal procedure with pipework is to follow through with a given diameter of pipework, picking up fittings, ancillaries and builder's work in the general order of flow. The type of pipe and jointing has been included in the heading, and so they do not require inclusion in the pipe item. The background for fixing pipe supports is classified as SMM General Rules 8.3. Painting to pipes is classified as SMM M60.9.0.2.0, and includes painting of pipe clips (SMM M60.C8).
Fittings to pipes ≤ 65 mm diameter are enumerated as extra over the pipes in which they occur, stating the number of ends (SMM Y10.2.3.4.0). Fittings which are reduced in size are measured the largest pipe in which they occur (SMM Y10.M5).

13·3

| 2/ | 1 | Pipe sleeve, len. ≤ 300, rigid PVC, to take 25 ⌀ steel pipe & hand to others for fixg.  &  Cut circ. hole thro. 50 th. in situ conc. duct cover for pipe ≤ 55 ⌀, m/gd. & fix only pipe sleeve.  &  Cut circ. hole thro. vinyl. asbestos flr. tilg. 3 th. for pipe ≤ 55 ⌀ & m/gd. |
|---|---|---|
| 2/2/ | 1 | Wall, flr. & clg. plates, 75 ⌀ × 6 th. c.p. zinc. alloy w. set screw fxg. to 25 ⌀ pipe. (flr. & clg. to f & r.) |
| 2/ | 2 | E.O. m.s. pipe, nom. size: 25, for fittg., 2 nr. ends. (bends) |

<u>Heavy quality black m.s. tube to BS 1387 w. welded jts. & fittgs. to BS 1965 (pipe supports m/s).</u>

Pipe sleeves are measured in accordance with SMM Y10.11.1.1.2. The supply of the pipe sleeves is likely to be a sub-contract item. However they are often fixed or built in by other trades. The fixing of the sleeves can conveniently be included with the builder's work item of cutting holes in concrete. Enumerated items for holes giving the nature of the structure, pipe size classification and other relevant particulars as SMM P31.20.2.1-3.2 & 4. The fixing only of pipe sleeves is covered by SMM P31.23.2.1.1. Wall, floor and ceiling plates are grouped together stating the type and size and method of fixing. (SMM Y10.12.0.1.0). It has been assumed that no sleeve is required through the ceiling and that the suspended ceiling is to be executed by a nominated sub-contractor. No hole is therefore measured through the ceiling.
Pipe fittings ≤ 65 mm diameter are described and enumerated as extra over the pipe in which they occur, stating the number of ends (SMM Y10.2.3.3.0). These will not be made bends.
It is necessary to take the pipe supports separately in this instance because the pipes are mounted one below the other; thus support of the lower pipe is taken from the higher one which in turn is supported by a drop rod from a channel section fixed to the soffit of the floor above. The channel will also support the ducting where appropriate. Additional information may be required from the engineer regarding the supporting brackets and steelwork to be provided.

13·4

131

| | | | |
|---|---|---|---|
| | MECHANICAL SERVICES (Contd.) | | The channel section which supports more than one installation will be billed in its own section (SMM P31.30.3.1.0). |
| 2/ | 3.10 | Pipes, st., nom. size 25. | Pipework is measured in metres stating the type, nominal size, method of jointing and supports (SMM Y10.1.1.1.0). |
| 2/ | 5 | E.O. m.s. pipe, nom. size: 25, for fittg. 2 nr ends. (bends | Number counted from isometric sketch (Bill Diagram Nr 1). |
| | 1 | Ditto., 3 nr ends. (tee | Descriptions as SMM Y10.2.3.3-4.0. |
| | | | A sketch (Bill Diagram Nr.2) is useful to illustrate what is measured and to assist the estimator. |
| | | BILL DIAGRAM Nr 2 Suppts. | |
| | 1 | Suppts. for pipes comprisg. bkt. inc. all nec. nuts, locknuts & washers to support two 25φ pipes, w. 2nr split rings, 10φ drop rod av. 200 lg., 75 x 50 x 6 m.s. chan. 150 lg. fxd. to conc. w. 12φ expandg. bolt as shown in Bill Diag. Nr 2. | Supports for services not provided with the services installation are enumerated and described in accordance with SMM P31.30.3.1.0. Alternatively, the similar provision in SMM Y10.9.0.1.3 could be considered more appropriate, when the size of the pipe is required to be stated. |
| | 1 | Ditto. comprisg. 2 nr split rings, 10φ drop rod. av. 200 lg. to stl. chan. (m/s). | Where the pipes are supported from a common channel bracket. |
| | | Combined component for supportg. pipes and ducting | A combined component to support both pipework and ducting which spans two work sections and could be covered by P31.30.3.1.0. (See also SMM Y10.9.0.1.3 and Y30.6.0.1.3). The spacing of the brackets will be governed by the smallest diameter service supported. In this case at 2.4 m centres. |
| | 1 | Suppts. for 25φ pipes & 300 x 300 ductg. comprisg. 75 x 50 m.s. chan. 1·20 m lg. fxd. to conc. w. 2 nr 12φ expanding bolts inc. all nec. nuts, locknuts & washers. | |

13.5

| | | | |
|---|---|---|---|
| | | Pipewk. ancillaries Valves | Pipework ancillaries are enumerated with a full description, reference to the project specification or the manufacturer's catalogue (SMM Y11.8.1.1.0). The jointing of the valves is given in the description of the item. The valves are provided with female taper thread ends. It is necessary therefore to include for nipples and unions as illustrated in Bill Diagram Nr 3. |
| | 2 | Pipewk. ancillary, valve, nom. size: 25, bronze wheel gate valve (Hattersley fig. 33X) inc. bronze nipple o/s & threadg. pipe o/s as Bill Diag. Nr 3. | |
| | | BILL DIAGRAM Nr 3 | |
| | 1 | Pipewk. ancillary, valve, nom. size: 25, bronze lockshield gate (ditto.) do. | The threading of the welded pipe could alternatively be measured separately as a special connection under SMM Y10.2.2.1.0. |
| | 1 | Pipewk. ancillary, valve, nom. size: 25, bronze 3 way motorized divertg. (Staefa Fig. M3P25G) inc. bronze nipple & union to each end. | The unions could alternatively be measured separately also. The following sketch will help to show the arrangement of the next valve. |
| | 1 | Pipewk. ancillary, valve, nom. size: 20, bronze wheel patt. strt. union radiator (Delflo fig. 2407) inc. bronze nipple & 25 - 20 reducer & threadg. pipe o/s & perforatg. convector casg. & Pipewk. ancillary, valve, nom. size: 20 ditto. lockshield patt. rad. (ditto.) do. | A study of the catalogue for the fan convector indicates that the connections are 20 mm BSP female threads and the fan convector casing requires perforation for the pipework. With the exception of radiator valves, identification discs are usually required for all control valves, engraved with the appropriate service (eg. HTG CWS). The motorized valve may also require a label affixed to it giving the valve number and the area served. These are enumerated in accordance with SMM Y54.3.2.1.0, stating the type, size and method of fixing. |
| | | Valve identification | |
| | 3 | Engraved ivorine valve identificatn. disc., 75 x 50, & fix to valve w. short len. of chain. | |

13.6

## MECHANICAL SERVICES (Contd.)

| | | |
|---|---|---|
| 1 | Engraved ivorine motorized valve identifcatn. label 100 × 50 & fix to valve w. adhesive. | A valve chart will also be required for the numbered valves in the plant room, indicating the duty, service and position of the valves. The chart would normally be mounted in a glass frame and fixed in the plant room; for which a separate enumerated item would be required (SMM Y54.3.6.1.2). |

Note: Fan convector FC4 is mesd. w. the ventilatn. ductwk. installatn. (m/s).

Where the fan convector is substantially part of the ductwork installation, it is usual to include it in that section. The insulation to the pipework where concealed in the floor trench (duct) and ceiling space is often kept separate in a bill of its own, as this work is usually a specialist sub-contract.

Sundries

| | | |
|---|---|---|
| Item | Mark the posns. of holes, mors. & chases in the structure for the LTHW htg. installatn. | SMM Y59.1.1.0.0 requires the marking out item to be given for each installation measured in the bill. |
| Item | Testg. & commsng. the LTHW htg. installatn. as descd. in the specfn. | A separate testing and commissioning item is required for each installation (SMM Y51.4.1.0.0). Reference to the specification avoids a lengthy description. |

13.7

## WARM AIR HEATING

In plant rooms

Ductg., galv. sheet stl. low velocity low pressure in accordance w. HVCA Specfn. DW/121 w. 'C' cleat cross jts. (fig. 41) & riveted spot or plug. welded lap longitudnl. seams (fig. 61). (suppts. m/s).

| | |
|---|---|
| 0·80 | Ductg., st., rect., 500 × 300 × 0·6 th. |
| 1 | E.O. 500 × 300 ductg. for fittg., 550 × 300 to 300 × 300 diminishg. piece. & Ducting ancillary, louvre, 550 × 300 alum. standard mill finish 'Neta-Line' type N/A 15 scrd. to tbr. & fxd. to 550 × 300 × 6 th. ductg. w. a flangd. and bolted jt. bedded in mastic sealant inc. flange on duct. |

Note: Tbr. fr. & opg. mesd. elsewhere.

The description follows that prescribed in SMM7-T40 (Appendix B). The ventilation ducting to the toilets would be classified under a heading of toilet extract (U 11: Appendix B).

Work within the plant room is identified separately (SMM Y30.M2).

The HVCA specification recommends various alternative forms of jointing. It is therefore necessary to specify the joints in the heading.

The ducting is measured over all fittings and branches (SMM Y30.1.1.1.0). The jointing is described in the heading and the fixing supports are measured separately. The thickness of metal varies depending on the width of the largest side. The ducting is deemed to include joints in running lengths and stiffeners where required (SMM Y30. C3).

Fittings are measured extra over the ducting in which they occur, stating the type (SMM Y30.2.3.1.0), and giving the larger size of ducting in the description for the diminishing piece. Louvres are enumerated as a ducting ancillary stating the type, size and method of jointing and fixing, and the size of ducting to which attached (SMM Y30.4.1.1.0). The nature of the builder's work detail must be ascertained from the engineer and/or architect. The louvre, strictly speaking, is external, but has been measured together with the ducting to which it directly relates.

The opening is part of the door frame to the plant room.

13.8

## MECHANICAL SERVICES (Contd.)

| | | | |
|---|---|---|---|
| | | len. (from louvre to wall face of plant rm.) | 2·400<br>200<br>2·600 |

| | | |
|---|---|---|
| 2·60 | Ductg., st., sq., 300 x 300 x 0·6 th. | Ducting measured as SMM Y30.1.1.1.0. |
| 1 | E.O. 300 x 300 ductg. for fittg., square bend. | Ducting fittings are described and enumerated as extra over the duct (SMM Y30.2.3.1.0). The items given in DW121 are used (HCVA Specification). The method of jointing is stated where different from the ducting in which the fitting occurs. Cutting and jointing ducts to fittings is deemed to be included (SMM Y30.C5). Air turns are enumerated separately where not provided with the fittings, stating the type and internal size of the ducting (SMM Y30.3.1.1.0). The maximum spacing for the air turns within the duct is 60 mm. |

60) 300
 5 spaces

| | | |
|---|---|---|
| 1 | Air turns, set of 4 aerofoil manufactured by Barber & Colman Ltd., to duct, int. size: 300 x 300. | |

60 mm max.

### Supports

| | | |
|---|---|---|
| 2/ 1 | Suppt. for 300 x 300 sq. ductg. of brackets inc. all nuts, locknuts & washers, comprisg. 25 x 3 m.s. flat lined w. 6 th. latex rubber, 2 nr. 10 ϕ drop rods 400 lg. & a 75 x 50 m.s. chan. 2ce fxd. to conc. w. 12 ϕ expandg. bolts. | It has been assumed that the supports are for the ducting only. The spacing and design of the supports is given in DW121. However, reference to the engineer is advisable to confirm the type of support required. To prevent noise transmission brackets in contact with the duct should be lined with sound absorbing material. The size of the ducting is given in the description in accordance with SMM Y30.6.0.1.1. It is important to make clear where the work in the plant room terminates. |
| | Note: Opg. in plant rm. wall mesd. elsewhere. | |

(End of work in plant room)

13·9

---

| | | |
|---|---|---|
| | len. (from plant rm. to FC4) | 1·600<br>1·400<br>2·500<br>5·500 |

Calculation of length of ducting in the Treatment Area in waste.

Galvd. sheet stl. low velocity low pressure rect. ductg. a.b.

| | | |
|---|---|---|
| 5·50 | Ductg., st., sq., 300 x 300 x 0·6 th. | The use of comprehensive headings reduces the length of the descriptions in the measured items that follow. Ducting measured as SMM Y30.1.1.1.0. |
| 2 | E.O. 300 x 300 ductg. for fittg., square bend. | Ducting fittings are measured as SMM Y30.2.3.1.0. |
| 2/ 1 | Air turns set of 4 nr aerofoil a.b. to duct, int. size: 300 x 300. | Air turns measured as SMM Y30.3.1.1.0. |
| 1 | E.O. 300 x 300 ductg. for fittg., stop end. | See SMM Y30.2.3.1.0. |
| 2 | Suppt. for 300 x 300 sq. ductg. of bracket comprisg. m.s. flat & 6 th. latex rubber, 2 nr. drop rods & chan. a.b. | See SMM Y30.6.0.1.1. Note method of abbreviating the description of an item previously described by the insertion of 'a.b.' (as before). |
| 2 | Ditto. comprisg. 25 x 3 m.s. flat & 6 th. latex rubber, 2 nr. 10 ϕ drop rods 400 lg. to stl. chan. (m/s). | The ducting is supported by a channel which also supports pipework. See SMM Y30.6.0.1.1. |

### Fan Convector

| | | |
|---|---|---|
| 1 | Fan coil heater convector 1518 x 705 x 254 hor. type (Dunham Bush series J42 Type 226) inc. type A manual air vent, single med. speed motor & 150 extension wall neck & 701 grille. | The fan convector is taken as an enumerated equipment item giving the type and limiting dimensions with a reference to the manufacturer's catalogue (SMM Y41.1.1.1.0). It is important to ascertain details of the connection to the pipework and ducting, supports and any associated builder's work. In this instance an extended wall neck is required to drop to the ceiling line. The joint to the ductwork need not be flexible in this instance. |

opening out in side of duct — concrete
300 x 300 duct — channels
access panel — drop rods each side
— wall neck
riveted and mastic sealant joint — ceiling — grille

BILL DIAGRAM Nr 4

13·10

| | | | | | |
|---|---|---|---|---|---|
| | MECHANICAL SERVICES (Contd.) | Plates, discs and labels for identification provided with the equipment are deemed to be included (SMM Y41.C3). The support bracket follows the design of the brackets previously measured, but the drop rods locate into fixing straps provided with the fan convector. The bill diagram will assist the estimator in calculating his rate for this and the following items. The support is measured in accordance with SMM Y41.6.1.0.0, stating the type, size and method of fixing. Connections between ducting and equipment are enumerated separately as SMM Y30.2.2.1.1, stating the type, size, ducting size and method of jointing. | | Insulation to LTHW heatg. installtn. Pipe insulatn. 19 th. of lg. fibred rockwool preformed sectns. & fittgs. secured w. lacing wire & covered w. canvas w. 3 nr. met. bands per sectn. | Particulars of the insulation are given as listed in SMM Y50.S1-5. It is measured in metres, stating the nominal size of the pipes as SMM Y50.1.1.1.0. In pipework of this size valve boxes are not required. Valve boxes are normally only required for the larger sizes in plant rooms. The insulation could merely stop at the valves, but it is probably better to take an extra over item for working the insulation around them, classified as ancillaries, as SMM Y50.2.1.1.0. |
| 1 | Suppt. to fan convector of bracket inc. all nuts, locknuts & washers, comprisg. 4 nr. 10 φ drop rods 200 lg. & 2 nr. 75 x 50 m.s. chan. sectns. 1·60 m lg. ea. 2œ fxd. to conc. w. 12 φ expandg. bolts, as shown in Bill Diag. Nr 4. | | 2/ 3·10 | 25 φ pipes. (LTHW Htg. to FC4 | |
| 1 | E.O. 300 x 300 ductg. for spec. conn., 1400 x 200, riveted & mastic sealed connectn. to fan convector inc. formg. opg. in duct. | | 4 | E.O. insulatn. to 25 φ pipes for workg. ard. ancillaries. (valves | Insulation is deemed to include smoothing the materials and working around supports; working around pipe flanges; and working around fittings (excluding metal clad facing insulants) as SMM Y50.C1. Hence there are no items required for working around bends and tees. |
| 1 | E.O. 300 x 300 x 0·6 th. ductg. for opg. & access dr., 1500 x 250, bolted & sealed. | Access doors are enumerated, giving the particulars listed in SMM Y30.2.4.1.1. The access door is required so that the connection can be made to FC4. Test holes and covers are enumerated and described as SMM Y30.2.6.1.1. The position of test holes must be clarified by the engineer. In this case 3 holes are required at the centre of the ducting run. Fire dampers are not shown in this run of ducting. However a query should be raised with the engineer as to whether a damper is required in the duct where it passes through the plant room wall. It is assumed not in this case. See SMM Y59.1.1.0.0. | | Loose Ancillaries | |
| 3 | Test hole & cover 25 φ to 0·6 th. ductg. & fit spring clipped cover (by F.T Products Ltd.). | | 6 | Loose keys to fit drain cocks & air vents & hand to employer's maintenance engineer. | Loose keys are enumerated separately stating the type and quality (SMM Y59.2.1.1.0). Clause 6 in the specification requires 6 loose keys to be provided. |
| | Sundries | | | Sundries | |
| Item | Mark the posn. of holes, mors., & chases in the structure for the warm air heatg. & toilet extract installtn. | | Item | Prepare, detailed dwgs. for bldrs. wk., 2 nr sets of 'as fitted' dwgs. & 1 nr set of photograph negatives, all as descd. in the specfn., & hand to the Engr. | Preparing drawings is inserted as an item, stating the information required, number of copies and details of negatives, prints and microfilm and name of recipient as SMM Y59.6.1.1.2. Drawings include builder's work, manufacturer's and installation drawings and record or 'as fitted' drawings (SMM Y59.D1). |
| Item | Testg. & commsng. the warm air heating & toilet extract installtn. as descd. in the specification. | See SMM Y51.4.1.0.0. The insulation work is classified in accordance with Sections R14-U70 in Appendix B of SMM 7. | | | |
| | 13·11 | | | 13·12 | |

# 9 ELECTRICAL SERVICES

## GENERAL BACKGROUND

The measurement of electrical services poses many of the same problems as the measurement of mechanical services. In fact the two are often grouped together and referred to as 'm and e services' as a collective term. It is unusual, however, for the work to be carried out as separate contracts and to be designed by separate specialist engineers. It is quite common for the surveying duties within a quantity surveying or consulting engineer's practice to be handled by separate specialist surveyors. There is a good deal of commonality in the approach to bill preparation.

A sound knowledge of electrical technology is required to understand the specification and to interpret the schematic drawings provided by the consulting engineer. A detailed knowledge of the IEE regulations for the electrical equipment of buildings and a knowledge of circuitry and wiring systems is essential so that trunking, tray and conduit runs can be plotted and the correct number of cables required measured for the two groups of services.

## MEASUREMENT PROCEDURES

The procedure for taking off electrical work is similar to that described for mechanical services (chapter 8) and therefore has not been repeated. A sound, systematic and logical approach with, possibly, the use of measurement schedules are the main requirements.

Where circuits are to be measured in detail, such as circuits other than lighting and small power, the route of the conduit and/or cable must be plotted on the plan or tracing overlay and the number of cables indicated. This sketch will then form a record of what is taken. An isometric sketch is often useful (as with pipework) to illustrate complex runs. Conduit and/or cable runs should be plotted using a standard nomenclature to illustrate high level, low level, rise, fall and number of cables. A suggested notation system is shown in the schedule overleaf.

When plotting conduit and cables it is usual to draw runs at right angles to each other rather than running diagonally. This is usually necessary because of the nature of the structure through which the conduits and cables are passing, as for example following joists or beams. Conduits and cables can sometimes be laid diagonally where running in floor screeds or in pitched roof spaces. Once the route has been plotted and the specification fully understood, the measurement is, as with mechanical services, relatively straightforward comprising basically enumerated items of equipment and final circuits and linear items of conduit, cable trunking, cable tray and cable, on more complex systems, all measured in accordance with the rules prescribed in SMM7.

Within the constraints of space in this book the worked example can only be an introduction to the subject although it introduces as many variations as possible. The main task facing the traditionally trained quantity surveyor is to develop his knowledge of technology as the measurement techniques involved are comparatively simple.

## DETAILED MEASUREMENT

### Conduit

Conduit (not in final circuits) is measured in metres, distinguishing between straight and curved, giving the radii, and stating the type, external size, method of fixing and background as SMM Y60.1.1–2.1.1–5, and particulars of materials as appropriate (SMM Y60.S1–6). The conduit is measured over all conduit fittings and branches (SMM Y60.M2). The conduit is deemed to include bending, cutting, screwing, jointing, and such conduit fittings as tees, elbows, bends, cover plates, bushes, locknuts,

*Standard notation to indicate route of conduit, tray or trunking*

Position

| | |
|---|---|
| In roof or above ceiling | — — — — — — — — |
| At high level | ——— · ——— · ——— |
| At low level | ———————— |
| Below floor | — — — — — — — |
| Vertical rise | ————⊙ F B |
| Vertical drop | ————⊗ T B |
| Change of level | —1.1—×—3.6— · — — |

Notation

| | |
|---|---|
| High level | H L |
| Low level | L L |
| From below | F B |
| To below | T B |
| From above | F A |
| To above | T A |

Number of cables in conduit or trunking

| | |
|---|---|
| 4 wires | ————//// ———— |
| 5 wires | ————//// ———— |
| 6 wires | ————//// / ———— |
| 6 wires (alternative method) | ————⑥———— |

nipples, stopping lugs and reducing bushes; clips, saddles and crampets; forming holes for conduit entry; draw wires and draw cables; and components for earth continuity (SMM Y60.C3 and Code of Procedure).

Special, adaptable, floor trap, purpose made and rectangular junction boxes and expansion joints are enumerated as extra over the conduit in which they occur, stating the type, size, cover and method of fixing, and background (SMM Y60.2.1–6.1.1). Cutting and jointing to conduit boxes is deemed to be included (SMM Y60.C4). Connections of conduit to trunking and to equipment and control gear are enumerated, stating the type, size and method of jointing (SMM Y60.3–4.1–2.1.0).

*Cable trunking, cable tray, ladders and racks*
These are measured similarly to conduit, but additionally stating the method of jointing and spacing and method of fixing supports (SMM Y60.5 & 8.1–2.1.1), and both trunking and cable tray are deemed to include components for earth continuity (SMM Y60.C5 & 7). Fittings of trunking, cable tray, ladders and racks, which include stop-ends, bends, tees, crosses, offsets and reducers, are enumerated as extra over the items in which they occur (SMM Y60.6 & 10.1.1.1 and Code of Procedure), while any necessary cutting and jointing to fittings is deemed to be included (SMM Y60.C6 & 8). Connections of cable trunking to equipment and control gear are enumerated, including forming holes and flanges, stating the size of opening and type and size of flange as appropriate (SMM Y60.7.1–3.1–2.1). Cable tray stools are enumerated stating the type and size (SMM Y60.9.1.0.0), while supports for trunking, cable tray, etc., which differ from those given with these items, are separately enumerated and fully described as SMM Y60.11–12.1.1.1.

*Cables*
Cables (not in final circuits) are measured in metres, giving the type, size, number of cores, armouring and sheathing, location, method of support and background as SMM Y61.1.1.1–7.1–2, and relevant particulars of materials as SMM Y61.S1–7. Cables in conduits or trunking, or fixed to trays, are measured as the net length of the conduit, trunking or tray, while other cables are measured as fixed without allowance for sag (SMM Y61.M2). The following allowances are made to cables which are measured net: (a) 0.30 m on each cable entering fittings, luminaires or accessories; and (b) 0.60 m on each cable entering equipment or control gear (SMM Y61.M3). Cables are deemed to include wall, floor and ceiling plates; cable sleeves; and connecting tails (SMM Y61.C3).

Flexible cable connections, cable joints, line taps, cable termination glands, and cable supports which differ from those given with cables are each enumerated, giving the various particulars prescribed in SMM Y61.2–6 inclusive. Busbar trunking, to carry three-phase supplies, is measured in metres, distinguishing between straight and curved, and giving the particulars listed in SMM Y61.7.1–2.1.1, while fittings such as stop-ends, bends, tees, crosses, offsets and reducers, are enumerated as extra over the busbar trunking in which they occur, stating the type (SMM Y61.8.1.1.0). Tap off units, feeder units, fire barriers, trunking supports and other ancillary items are enumerated, giving the particulars prescribed in SMM Y61.9–12 & 14–18.

*Cable and conduit in final circuits*
These are enumerated on an enumerated points basis where they form part of a domestic or similar simple installation from distribution boards and the like (SMM Y61.M7), as in the installation illustrated in example 14, where there is a reasonably consistent relationship between the length of conduit and cable and the number of points

that they serve. The Code of Procedure points out that this approach is appropriate for the majority of small power and lighting installations of a domestic or similar nature and also to the more simple installations in final circuits in other sections. Other types of final circuit are measured in detail in accordance with SMM Sections Y60, Y63, Y61, Y62 and Y80 (SMM Y61.M6).

SMM Y61.P2 requires the following information to be given in connection with final circuits:

(a) a distribution sheet setting out the number and location of all fittings and accessories; and
(b) a location drawing showing the layout of the points.

Examples of both of these documents are illustrated on Drawing 14 in this chapter. The Code of Procedure recommends that the distribution sheet should contain information relating to the location, number and type of lamps, the number of lighting, switch and socket points and the type of fittings, appliances and accessories, together with any other information relevant to the circuit arrangement for each distribution board and the like.

Enumerated items for cable and conduit in final circuits are to state the size and type of cable/conduit and give a description of the final circuit, including the number of sockets, switch outlets and the like; immersion heaters, cooker outlets and the like; lighting outlets; one way switches; two way switches; and intermediate switches; together with earthing details where an integral part of the final circuit, special boxes, whether surface or concealed, and background and method of fixing (SMM Y61.19.1–2.1–6.1–5), and also the voltage and amperage (SMM Y61.S8).They are deemed to include conduit accessories, including conduit boxes required for the particular installation; fixing, bending, cutting, screwing and jointing; and determining routes (SMM Y61.C6). Each lighting outlet is measured as one point irrespective of the number of lamps (SMM Y61.M8).

Further clarification of the measurement of enumerated final circuit items is given in the Code of Procedure and includes the following guidelines:

(1) It is not necessary to give the size of conduit as it will be at the discretion of the contractor.
(2) The 'distribution boards and the like' from which final circuits are measured include such control gear as control panels for boilers, fire alarms, or master clocks and similar items.
(3) The classification of points in the enumeration of final circuits relates to the terminations of the permanent wiring to switches and to outlet accessories and control gear for the connection of current using appliances or fittings.
(4) Where final circuits are connected to multi-gang accessories, the number of points will normally be the same as the number of gangs. Where the gangs are electrically wholly interconnected within the accessory, such interconnected gangs should count as one point.
(5) Flexible conduits, cables and the like between appliances or fittings and the associated terminal accessories or control gear on the permanent wiring of a final circuit, should be included in the description of the relevant appliance or fitting.
(6) Examples of final circuits of different types, which should be identified in the circuit descriptions, include single outlet radial circuits, multiple outlet radial circuits, ring circuits, circuits wired in series and open circuits.

*Switchgear and distribution boards*
These are enumerated, stating the type, size, rated capacity and method of fixing, usually with a cross reference to the specification, and giving details of fuses, supports provided with the equipment and method of fixing, and background as SMM Y71.1–2.1.1.1–3. Supports not provided with the equipment are enumerated and described separately (SMM Y71.5.1.0.1). Providing everything necessary for jointing is deemed to be included (SMM Y71.C1), as also are plates, discs and labels for identification provided with the equipment (SMM Y71.C3). The Code of Procedure states that fuse links and miniature circuit breakers supplied with the switchgear and distribution boards shall be included in the description of the control gear; whereas those supplied independently shall be measured separately.

*Luminaires and accessories for electrical services*
These are covered in SMM sections Y73 and Y74. Luminaires are enumerated stating the type, size and method of fixing, often with a cross reference to the specification, and any other relevant details listed in the fourth column (SMM Y73.2.1.1.1–13). Pendants are also enumerated, distinguishing between those with a drop $\leq 1.00$ m and those $> 1.00$ m. Lamps may be separately enumerated, stating the type, size and rated capacity (SMM Y73.3.1.0.0), or alternatively they can be given in the description of the luminaires (SMM Y73.M2). Accessories, which include lighting switches, socket outlets, thermostats, telephone cord outlet points and bell pushes, are enumerated stating the type, box, method of fixing, background and rated

capacity, and may provide for plugs to be provided with socket outlets (SMM Y74.5.1.1.1–2). The description of accessories should state the number of gangs comprised in the accessory.

There is also provision for enumerated particular specification items in SMM section Y74, which are to include the type and description (SMM Y74.1.1.0.1–13). They are classified as items of a fitting or ancillary nature (SMM Y74.D2). Further clarification is given in the Code of Procedure, where examples of particular specification items include clocks, telephones, alarm bells, loud speakers, battery chargers, convector heaters, storage heaters, telephone equipment, facsimile equipment, aerials, microphones, amplifiers, recording/playback equipment, central clock control equipment, computer control systems, alarm equipment, security alarm equipment and control indicator panels.

*Testing and commissioning electrical services*
This is given as an item, stating the installation and any other relevant requirements listed in SMM Y81.5.1.1–2.1–2. Provision of electricity and other supplies and of test certificates is deemed to be included (SMM Y81.C1 & C2).

*Identification of electrical work*
This is enumerated, where not provided with equipment or control gear, categorised as to plates, discs, labels, tapes and bands, arrows, symbols, and numbers, and charts, giving the type, size and method of fixing (SMM Y82.4.1–6.1.1–2).

*Sundry items*
These are covered in SMM section Y89. For example, marking position of holes, mortices and chases in the structure is given as an item, stating the installation as SMM Y89.2.1.0.1. Loose ancillaries, such as keys, tools and spares are enumerated, stating the type, quality or quantity and name of recipient (SMM Y89.3.1–3.1.1). Another item could comprise the temporary operation of installations to the employer's requirements, stating the installation and purpose and duration of operation, attendance required, and any conditions and special insurance requirements of the employer (SMM Y89.6.1.1.1–3). Provision of electricity and other supplies required for the temporary operation of the installation is covered by Provisional Sums as listed in SMM A54.

Preparing drawings is included as an item, giving the information required and number of copies, and details of negatives, prints and/or microfilms, together with details of binding into sets where required and names of recipients (SMM Y89.7.1.1.1–2). Drawings include builder's work, installation drawings and record or 'as fitted' drawings (SMM Y89.D1). A further item could relate to operating and maintenance manuals where specified (SMM Y89.8.0.0.0).

## WORKED EXAMPLE

A worked example shows the method of measuring the electrical services to a school extension in accordance with the details shown on drawing 14 and the following specification.

The student is advised to work carefully through this example taking note of the comments accompanying the taking off and referring to the appropriate clauses of SMM7.

## SPECIFICATION NOTES FOR ELECTRICAL INSTALLATION TO SCHOOL EXTENSION

*Generally* — Voltage system — single-phase 240 volt, 50 Hz. Wiring system to be 600/1000 volt grade PVC insulated and colour-coded cables drawn into heavy gauge mild steel conduit. Conduit to be concealed in plastered walls or sunk in screeded floors or surface mounted in ceiling void. Conduit to have screwed fittings and to act as earth continuity throughout.

*Lighting circuits* 1 mm$^2$ single-core PVC insulated cables in 3 circuits, in generally 20 mm diameter conduit, mainly located in ceiling void.

*Power circuits* 2.5 mm$^2$ single-core PVC insulated cables in two ring circuits in 20 mm diameter conduit, mainly embedded in screeded floors.

*Fixed apparatus circuit* 6 mm$^2$ single-core PVC insulated cables in 20 mm diameter conduit routed along corridor ceiling, ceiling of room 1/102 and drop to outlet.

*Distribution board* Surface type metal clad consumer unit with 100 amp main switch and 8 ways for miniature circuit breakers 3 at 5 amp,

| | |
|---|---|
| | 3 at 30 amp and 2 spare. Model KM 100 or similar. |
| Main cables | 35 mm² single-core PVC insulated cables in connection from existing 3-phase distribution board, laid into existing trunking in existing corridor and drawn into new 32 mm diameter conduit to new distribution board '1/12'.<br>Line and neutral conductors to be led into existing distribution board and connected to one single-phase way and include supply of 100 amp HRC cartridge fuse.<br>Earthing continuity of new system to be provided through new conduit by fixing to trunking with locknut and star washer connection. |
| Pendant fittings | 3-core, 0.75 mm² PVC insulated and protected flexible 800 mm drop, ivory plastics ceiling rose and brass BC lamp holder with shade ring. |
| Soffit fitting | Opalescent bowl fitting catalogue reference UK 275, with two ceramic BC lamp holders. |
| Bulkhead fittings | Polycarbonate bowl fitting, catalogue UK 155, with one ceramic BC lamp holder, mounted on external wall. |
| Fluorescent fittings | 1500 mm long 65 watt single tube pattern with diffuser, catalogue reference UK 654. |
| Lighting switches | 5 amp single-pole silent action ivory plastics plate switches, one-way, two-way or in gangs as appropriate. Switches to be 1400 mm above floor level. |
| Power outlets | 13amp switched socket outlets, flush ivory plastics pattern in single or double units. Earth continuity to be provided between conduit box and socket outlets with 1 mm² sleeved copper cable. Power outlets to be 450 mm above floor level. |
| Fixed apparatus outlet | Surface metal clad pattern 30 amp switched outlet with screwed outlet for 20 mm flexible conduit of reprographic apparatus. Outlet to be 600 mm above floor level. (Flexible conduit and final connections to apparatus are not part of this contract.) |
| Plans and label | Contractor to prepare and provide 4 prints of plans of installation 'as fitted' showing circuits and conduit runs. Contractor to provide white plastics label 'DIS BOARD 1/12' with list of each circuit under. Label to be fixed to wall adjacent to consumer unit. |

# DRG. 14 ELECTRICAL SERVICES TO SCHOOL EXTENSION

## INTERNAL ELEVATION - EXISTING CORRIDOR
Scale 1:100

- existing 3 phase distribution board
- existing 100 x 50 m.s. trunking on surface
- new connection for proposed extension in 32 m.s. heavy gauge conduit
- new opening formed in outer wall for proposed extension

## SECTION A-A
Scale 1:100

- proprietary suspended ceiling system
- timber roof construction
- new 32 diam. m.s. heavy gauge conduit on surface inside cupboards
- surface metal clad consumer unit (Distribution Board 1/12)

## DISTRIBUTION SHEET
Distribution Board :- 1/12 (all single phase - 240V - 50Hz)

| Location | Lighting Circuit Nr. | Fittings | Switches | Power Circuit Nr. | 13A Single SSO | 13A Double SSO | Special Power Circuit Nr. | | Remarks |
|---|---|---|---|---|---|---|---|---|---|
| Room 1/100 | 1 | 15 65W flu. tubes | 2 x 3 gang 2 way | 4 | 2 | 5 | - | - | 65W fluorescent single tube ref: UK 654 with diffuser |
| Room 1/101 | 2 | 6 65W flu. tubes | 1 x 2 gang 1 way | 5 | 2 | 3 | - | - | |
| Room 1/102 | 2 | 4 plain pendant 150W lamps | 1 x 1 way | 5 | 2 | 1 | 6 | 1 x 30A | special power to photo-copier |
| Corridor/ Vestibule | 3 | 6 65W flu. tubes | 1 x 2 gang 2 way | 5 | 2 | - | - | | |
| | | 1 soffit mounting | 1 x 2 gang (1) 2 way (1) 1 way | | | | | | soffit mounting fitting opalescent bowl. 2 B.C. lamps ref: UK 275 |
| | | 2 bulkhead (outdoor) | 1 x 2 way | | | | | | bulkhead fittings B.C. lamps ref: UK 155 |
| Totals | | 34 | | | 8 | 9 | | 1 | |

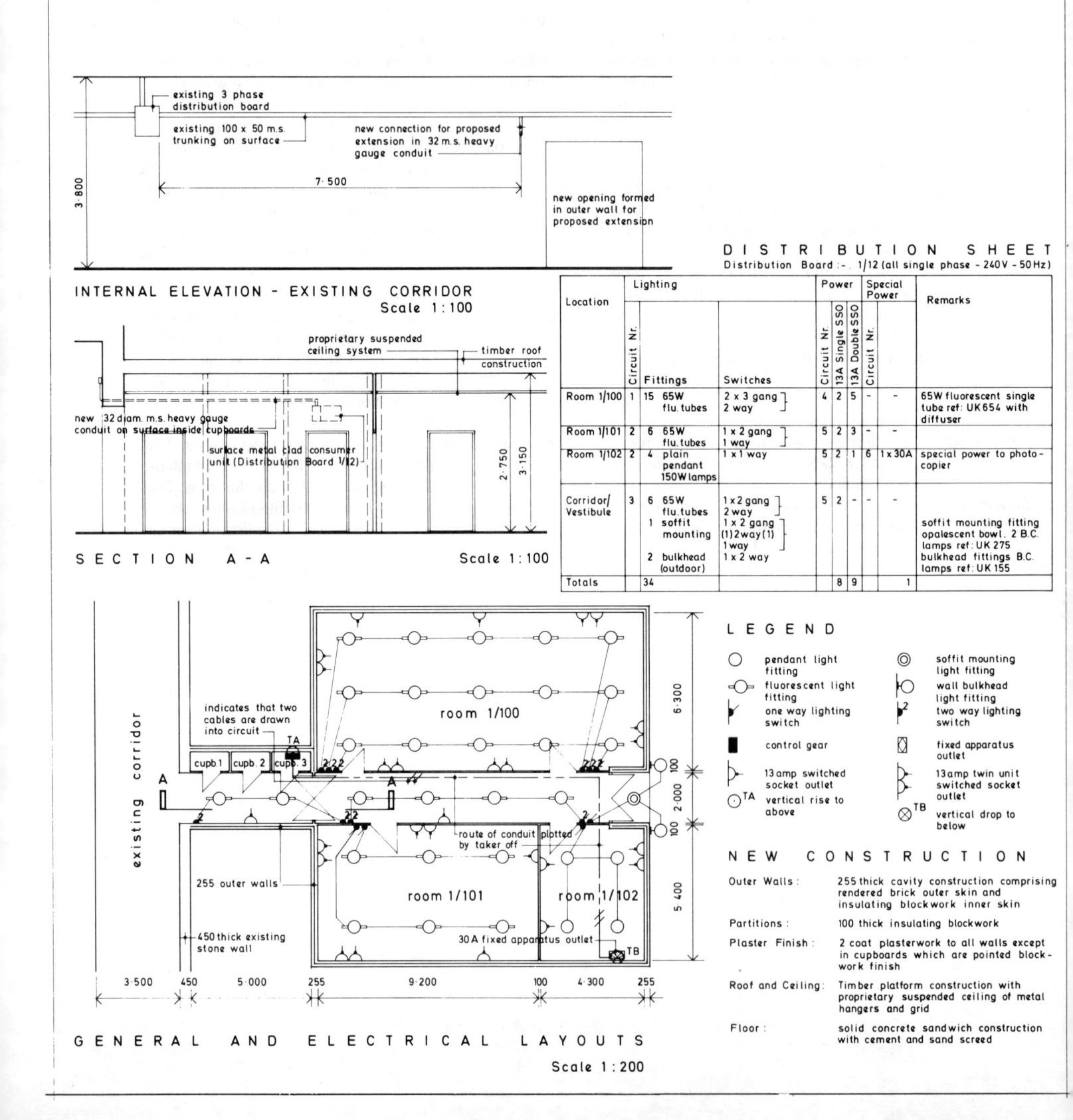

## GENERAL AND ELECTRICAL LAYOUTS
Scale 1:200

- existing corridor
- cupb.1, cupb.2, cupb.3
- room 1/100
- room 1/101
- room 1/102
- 255 outer walls
- 450 thick existing stone wall
- 30A fixed apparatus outlet
- indicates that two cables are drawn into circuit
- route of conduit plotted by taker off

## LEGEND

- pendant light fitting
- soffit mounting light fitting
- fluorescent light fitting
- wall bulkhead light fitting
- one way lighting switch
- two way lighting switch
- control gear
- fixed apparatus outlet
- 13 amp switched socket outlet
- 13 amp twin unit switched socket outlet
- TA vertical rise to above
- TB vertical drop to below

## NEW CONSTRUCTION

**Outer Walls:** 255 thick cavity construction comprising rendered brick outer skin and insulating blockwork inner skin

**Partitions:** 100 thick insulating blockwork

**Plaster Finish:** 2 coat plasterwork to all walls except in cupboards which are pointed blockwork finish

**Roof and Ceiling:** Timber platform construction with proprietary suspended ceiling of metal hangers and grid

**Floor:** solid concrete sandwich construction with cement and sand screed

# ELECTRICAL SERVICES TO SCHOOL EXTENSION

### Information

Drawing Nr 14 and Specification Notes indicate the scope of the work.

### Description of the installations

The wks. comprise the complete new electrical installatn. to the proposed school extsn. The mains are brought from an xtg. distributn. bd. situated in the corridor adjoing. the extsn. & run to a new bd. in cupbd. Nr 3. The installations comprise both lighting and power. The wiring system consists of 600/1000 volt grade PVC insulated & colour coded cables drawn into heavy gauge m.s. conduit.

### Regulations, rules & byelaws

The wk. shall be carried out in accordance with the IEE Regulations for the Electrical Equipt. of Bldgs. & the rules of the Supply Authority.

### Main Supply

The 240 volt 50 Hz main supply is taken from an xtg. 3 phase distributn. bd. located in the corridor adjoing. the new school extsn.

### Mains Installatn.

|  | Main supply from xtg. corridor to DB/12 |
|---|---|
|  | Conduit |
|  | new bldg. |
|  | 4·000 |
|  | 100 |
|  | 4·100 |
|  | xtg. bldg. |
|  | 300 |
|  | 500 |
|  | 800 |

---

Detailed project specifications of electrical installations together with drawings indicating the scope and location of the work will assist the quantity surveyor in measuring the work and the estimator in pricing it. A general description of the installations is not required by SMM7, but it could prove helpful to the estimator. The location drawings do not always show the electrical installation in adequate detail. It could be argued that as a location drawing and distribution sheet are required by SMM Y61.P2, that this brief description is superfluous. It is however useful to show the student the method of description preparation. Two important preamble clauses have been inserted for the benefit of the student. Reference to the consulting services engineer may be needed to ascertain if the installations are to comply with any special regulations.

The type and voltage of the main electricity supply and the location of the main supply point have been included for information purposes.
The sequence of taking off follows the order given in SMM 7.
Provide ample headings and sub-headings to act as signposts throughout the dimensions.
The conduit can be scaled directly from the drawing.

14·1

---

# EXAMPLE 14

| | | |
|---|---|---|
| | 4·10 | Conduit, strt., 32 ⌀ black enamelled heavy gauge m.s. fixd. to surfs. w. crampets to masonry background. |
| | 0·80 | (in xtg.bldg.) |
| | 1 | Conn. of conduit, 32 ⌀, to xtg. cable trunkg., inc. formg. hole & locknut & starwasher. |
| | | Cables |
| | | 4·100 |
| | | tail 600 |
| | | 4·700 |
| 2/ | 4·70 | Cable 35 mm² single-core PVC insulated 600/1000 volt grade colour coded, drawn into conduit. |
| 2/ | 0·80 | (in xtg. bldg.) |
| | | 7·500 |
| | | tail 600 |
| | | 8·100 |
| 2/ | 8·10 | Cable, 32 mm² ditto., drawn into xtg. 100 × 50 m.s. trunkg. |
| | 1 | Fuse 100 amp HRC cartridge, & fit to spare way in xtg. distributn. bd. |

This is unusual as in practice the route of the conduit would need to be drawn on the plan or on a tracing paper overlay. (See the measurement of conduit and cable to the fixed apparatus outlet).
The description of conduit shall state whether straight or curved, type, external size, method of fixing and nature of background (SMM Y60.1.1.1.1-2). Conduit fittings are deemed to be included with the conduit item (SMM Y60.C3 and Code of Procedure).
This is not a separate item within the scope of SMM7 (electrical services - work to existing buildings).
The locknut and starwasher connection to the trunking is necessary to maintain earth continuity. This forms an enumerated item in accordance with SMM Y60.3.0.1.0.
Cables are measured in metres stating the type and size, number of cores and other relevant particulars as SMM Y61.1.1.1.0. Allowance has been made for cable tails at both ends at each distribution board in accordance with SMM Y61.M3. The removal and replacement of the existing trunking cover is deemed to be included (SMM Additional Rules: Y.C2).
It is necessary to state the type and size of the existing trunking (SMM Additional Rules: Y.2.1.2.0).

Fuse links (to hold fuse elements) supplied independently of control gear are measured separately (Code of Practice Y71).
An HRC cartridge fuse is a high rupture capacity fuse filled with silica sand.

14·2

## ELECTRICAL SERVICES (Contd.)

| | | | |
|---|---|---|---|
| 1 | Distributn. bd. 8 way surf. type metal clad m.c.b. consumer unit w. 100 amp main switch, manufacturer's ref. KM.100, w. three 5 amp & three 30 amp m.c.b.s. & plugd. & scrd. to masonry. | | M.c.b. is an abbreviation for miniature circuit breaker. The specification states that the distribution board is to be model KM 100 or similar. To maintain parity of tendering the option has been excluded from this item. The successful contractor will be given the opportunity to suggest alternative manufacturers who may be considered after the tender has been let. The description of the distribution board is to include the type, size, rated capacity, fuses, method of fixing and background as SMM Y71.2.1.1.1-3. The background classification conforms to SMM General Rules 8.3(b). The two sundries items are required to be given for each installation. The item for marking the position of holes is covered by SMM 89.2.1.0.0, whereby the electrical sub-contractor marks out for the main contractor's builder's work. |
| | Sundries | | |
| Item | Markg. posn. of holes, mors, & chases in the struct. for the mains installatn. | | |
| Item | Testing & commsng. the mains installatn. as descd. in the specfn. | | See SMM Y89.5.1.1-2.1-2. Additional particulars may be included in the item description where considered necessary. Cables and conductors for earthing not forming an integral part of the circuit are measured in detail. Accessory measured as SMM Y74.5.1.1.2. The particulars are extracted from the specification notes. This switched outlet serves the fixed apparatus (photo-copier). Code of Practice: Y61.19 indicates that enumeration on a points basis of work in final circuits is appropriate for small power and lighting install-ations of a domestic or similar nature to the more simple installations in other sections. |
| | Accessory | | |
| 1 | Soc. outlet 30 amp (Rm. 1/102 surf. met. clad patt., switched, w. scrd. 20φ outlet & k.o. box, plugd. & scrd. to masonry. | | |

14.3

---

Conduit & cable to fxd. apparatus outlet

to walls
rise from d.b.  650
drop to outlet
            3·150
less   600   2·550
            3·200

to tbr. clg.
run in clg.    1·000
from d.b. to Rm. 1/102  13·000
            7·500
            500
           22·000

| | | |
|---|---|---|
| 3·20 | Conduit, strt., 20φ black enamelled heavy gauge m.s., fxd. to surfs. w. crampets to masonry backgrd. | (rise to d.b. & drop to outlet |
| 22·00 | Ditto. 20φ do. fxd. to surfs. w. clips to met. backgrd. | (run in clg. from d.b. to Rm. 1/102 |
| 2/ 3·20 | Cable, 6 mm² single-core PVC insulated 600/1000 volt grade colour coded drawn into conduit. | (rise & drop (run in clg. |
| 2/ 22·00 | | |
| 2/ 0·30 | | (tails at outlet |
| 2/ 0·60 | | (tails at d.b. |

It has been decided that this particular conduit and cable do not come within its scope and are accordingly measured in detail. Some surveyors might not agree with this interpretation. The selected approach also has the advantage of illustrating an alternative method of measurement to the student.
The conduit has been plotted on the drawing using the standard notation illustrated in the text of the chapter.
In practice the experienced taker off would not use such detailed side casts.
The scale rule would normally be used to determine running dimensions. This follows the technique used by electrical estimators. It is vital however for each dimension given in the taking off to be adequately signposted.
By virtue of SMM Y60.C3(a) the conduit item is deemed to include all conduit fittings and labours, except special boxes as listed in SMM Y60.2.1.-5.
The conduit is classified in accordance with SMM Y60.1.1.1 1-2, giving the background to which the conduit is fixed. In the ceiling void it is clipped to metal channels supporting the proprietary ceiling.
It is not necessary to separate conduit fixed at different levels. The measurement of cables in conduit follows the dimensions of the conduit (SMM Y61.M2). Allowance must be made for cable tails, 0·30 m at fittings and accessories, and 0·60 m at equipment and control gear. The cable dimensions must be timsed by the number of cables running in each conduit shown on the layout plan.

14.4

ELECTRICAL SERVICES (Contd.)

### Power Installatn.

#### Final circuits

| | |
|---|---|
| 1 | Cable & conduit in final ring main circuit, concealed installatn. consistg. of 2.5 mm² PVC insulated colour coded cable drawn into black enamelled heavy gauge m.s. conduit, embedded in screeded flrs., in circuit nr 4 comprisg. seven s.s.o's. |
| 1 | Ditto. in circuit nr 5, comprisg. ten s.s.o's. |

#### Accessories

| | | |
|---|---|---|
| 2 | Soc. outlet, 13amp 3 pin shuttered flush ivory plastic patt. single switched, & steel k.o. box, plugd. & scrd. to masonry. | (Rm. 1/100) |
| 2 | | (Rm. 1/101) |
| 2 | | (Rm. 1/102) |
| 2 | | (Corridor/Vestibule) |
| 5 | Ditto. double switchd. do. | (Rm. 1/100) |
| 3 | | (Rm. 1/101) |
| 1 | | (Rm. 1/102) |

14.5

Colour coded cables must be so described (SMM Y61.S7). Cables are classified in accordance with SMM Y61.1.1.1.0. Final circuits are enumerated and it is good practice to give the circuit reference in each case, extracted from the distribution sheet for identification puposes, although not specifically required by SMM Y61.19. Only where final circuits are identical in every respect would they be measured together. An example of this would occur in a multi-storey building with identical facilities provided on each floor. It is necessary to indicate whether the power circuit is a radial or ring main circuit and the number of sockets.
It is not necessary to give the size of conduit as it will be at the discretion of the contractor (Code of Procedure : Y61.19). It is necessary to state whether the conduit/cable is attached to a surface or concealed and the nature of the background and method of fixing (SMM Y61.19.2.1.3-5). Accessories are enumerated and the descriptions generally refer to the specification and contain details of the type, box, method of fixing and background as listed in SMM Y74.5.1.1.0. Accessories include switches and socket outlets (Code of Practice : Y74.5). The conduit box to which the socket outlet is fixed is included in the description as a connection item. The abbreviation 'k.o. box' refers to a knock out box. Alternatively, it might be deemed to be included in the conduit in the final circuit.

#### Sundries

| | |
|---|---|
| Item | Markg. posn. of holes, mors. & chases in the struct. for power installatn. |
| Item | Testg. & commsng. the power installatn. as descd. in the specfn. |

### Lighting Installation

#### Final circuits

| | |
|---|---|
| 1 | Cable & conduit in final circuit, concealed installatn. consistg. of 1.00 mm² PVC insulated colour coded cable drawn into black enamelled heavy gauge m.s. conduit, fxd. to met. surfs. w. clips, in circuit nr 1, comprisg. fifteen lighting points & six 2 way switch points. |
| 1 | Ditto. circuit nr 2 comprisg. ten lighting points & three 1 way switched points. |
| 1 | Ditto. circuit nr 3 comprisg. nine lighting points & one 1 way & four 2 way switch points. |

#### Luminaires

| | | |
|---|---|---|
| 4 | Luminaire, pendant fittg. comprisg. ivory plastic clg. rose & connector blk. & brass B.C. lampholder w. shade ring & 3-core 0.75 mm² PVC insulated & protected flexible cable, drop ≤ 1.00 m, inc. 50 ⌀ loop in conduit box. | (Rm. 1/102) |

14.6

The earth continuity fixing described in the specification is included in the conduit in accordance with SMM Y60.C3(e).
Sundries items as before are required for each installation (SMM Y89.2.1.0.0).

See SMM Y89.5.1.1-2.1-2.

All final circuits that are measured on an enumerated basis are shown on a distribution sheet. The distribution sheet sets out the number and location of all fittings and accessories, which include luminaires, lamps and lighting switches.
The distribution sheet provided with this example is adequate for the purposes of SMM Y61.P2(a) and must be inserted in the bill of quantities.
The method of measurement of final circuits is prescribed in SMM Y61.19.2.1-6.1-5.

The connector block mentioned in the description is a standard inclusion with most pendant fittings.
However SMM Y73.2.2.1.2,4, 5 & 9 requires the item to be given in the description.

144

| | | ELECTRICAL SERVICES (Contd.) | | Where fittings are fixed flush with a suspended ceiling it is usual to provide a loop in the conduit box (with openings for conduit in the back of the box) flush with the ceiling to which the fitting would be screwed. This box is included with the item. There is no requirement to keep external electrical work separate in SMM 7. Descriptions can be reduced in length by referring to specific clauses in the project specification. This is an external wall fitting wired into an adjoining lighting circuit. Although not mentioned in the specification, the conduit box to which the fitting would be fixed would need to be galvanised. The description is to include the method of fixing. Fluorescent fittings require two fixing points. One would be provided with the conduit installation and the other must be included in the description or alternatively measured separately. The bracket is made up in conduit and screwed to the timber roof. The support bracket may not coincide with a joist. To measure the bearers reference would need to be made to the roof drawing. Lamps are enumerated, stating the type, size and rated capacity (SMM Y73.3.1.0.0). It is common practice to take the lamps separately from the luminaires, although they may alternatively be given in the description of the luminaires (SMM Y73. M2). | | 2 | Lamp, 60W B.C. G.L.S as specfd. (vestibule | The wattages of the lamps to the vestibule and the external bulkhead fittings are not given in the specification and have been assumed. Alternatively these could form architect's queries. Lighting switches are classified as accessories in Code of Procedure: Y74.5, and the descriptions are to conform to SMM Y74.5.1.1.2. The descriptions of switches are to give the number of gangs in accordance with SMM Y74. M3. The knock out box is included in the description. Proceed logically through the lighting switches, checking against the layout drawing and the distribution sheet. Always double check the quantities to ensure that they are correct. |
| | 1 | Luminaire, opalescent bowl fittg. cat. ref. UK 275 w. connector blk. & 2 nr ceramic B.C. lampholders, ditto. (vestibule | | 2 | Ditto. 100W. (Bulkhd. ext. | |
| | | | | 27 | Lamp, fluorescent tube, 1·50 m lg. 65W 'warm white'. Accessories | |
| | 2 | Luminaire, polycarbonate bowl fittg. cat. ref. UK 155 w. connector blk. & one ceramic B.C. lampholder, inc. 50 ø galvd. m.s. loop in galvd. conduit box, plugd. & scrd. to masonry. | | 1 | Lightg. switch 1 gang 1 way 5 amp single pole silent action, ivory plastic plate & stl. k.o. box, plugd & scrd. to masonry. (Rm. 1/102 | |
| | | | | 1 | Ditto. 1 gang 2 way. (vestibule | |
| | 27 | Luminaire, fluorescent fittg. 1·50 m lg. 65W single tube & diffuser cat. ref. U.K.654, inc. conn. blk. 50 ø loop in conduit box & one support bkt. comprisg. 600 len. of 20 ø conduit & 2nr 50 loop in boxes; one box scrd. to tbr. | | 1 | Ditto. 2 gang 1 way. (Rm. 1/101 | |
| | | | | 1 | Ditto. 2 gang 2 way. (corridor | |
| | | | | 1 | Ditto. 2 gang (1) 1 way (1) 2 way. (vestibule | |
| To take | | | | 2 | Ditto. 3 gang 2 way. (Rm 1/100 | |
| Bldr's. wk. — tbr. brrs. spanng. rf. jsts. to take fixg. bkt. for fluorescent fittings. | | | | | Sundries | |
| | | | | Item | Markg. posn. of holes, mors, & chases in the struct. for the lighting installatn. | Sundries item as required for each installation. See SMM Y89.2.1.0.0. |
| | | Lamps | | Item | Testg. & commsng. the lighting installatn. descd. in the specfn. | See SMM Y89.5.1.1-2.1-2. |
| | 4 | Lamp, 150 W B.C. G.L.S. as specfd. (Rm. 1/102 | | | Identificatn. items | |
| | | | | 1 | Identificatn. label, 125 x 100, white plastic, marked DIS BOARD 1/12 w. list of ea. circuit beneath, plugd. & scrd. to masonry. (sited near consumer unit | The identification label is not provided with the control gear and so is required to be measured under SMM Y82.4. 3.1.0, stating the type, size and method of fixing. |

14·7      14·8

| | | | | |
|---|---|---|---|---|
| | ELECTRICAL SERVICES (Contd.) | | | |
| | *Sundries* | | | |
| Item | Preparg. dwgs., 4 copies of 'as fitted' prints showg. circuits & conduit runs, & hand to maintenance engr. | See SMM Y89.7.1.1.2. One item only required for the whole works. It is a common requirement for these drawings to be on linen. If this were the case here then this would need to be stated in the description. Description is to include number of copies and name of recipient. | | |
| | *Builder's work* | | | |
| 1 | Cuttg. hole for circ. pipe ≤ 55 nom. size thro. xtg. 450 th. stone wall plastered 1 side & m/gd. pla. to match xtg. | Cutting hole in existing building for conduit is measured in accordance with SMM Additional Rules 6.2.1.2 & 5. | | |
| 0.30 | Paintg. servs., isoltd. surfs., gth. ≤ 300, wirebrush & pt. ① calc. plumbate primer, ① u/c & ② gloss pt. (in xtg. bldg.). | It has been assumed that the trunking does not require painting. The painting specification has been assumed and follows common practice. Measured in accordance with SMM M60.9.0.2.0. Services include conduit for painting purposes (SMM M60. D12). | | |
| | *Cuttg. or formg. holes, mors., sinkgs. & chases for electrical installtns. & m/gd.* | The quantities are taken from the distribution sheet and previous dimensions. | | |
| 1 | Exposed m.s. conduit, 32 φ, 1nr. equipt. & control gear pt. (mains supply | The cutting or forming holes, mortices, sinkings and chases for electrical installations are enumerated, distinguishing between concealed and exposed conduit/cable, stating the type and giving the number of classified points and including making good as SMM P31.19.1-4.1-5.1-2. | | |
| 1 | Concealed m.s. conduit, 20 φ, 1 nr fittg. outlet pt. (fxd. applce. | | | |
| 1 | Concealed m.s. conduit, 20 φ, 7 nr soc. outlet pts. (power circt. nr 4 | | | |
| 1 | Concealed m.s. conduit, 20φ, 10 nr soc. outlet pts. (power circt. nr 5 | | The points are enumerated irrespective of the type, size and kind (SMM P31. M10), and associated switch points are deemed to be included (SMM P31. C3). No distinction is made between single and double switched socket outlets. | |
| 1 | Concealed m.s. conduit, 20 φ, 15 nr luminaire pts. (lightg. circt. nr 1 | | | |
| 1 | Ditto., 10 nr do. (do. nr 2 | | | |
| 1 | Ditto., 9 nr do. (do. nr 3 | | | |

14.9        14.10

# 10 ALTERATION WORK AND REVISION NOTES

## EXTENT OF WORK

Alteration work comprises the demolition of existing buildings, alterations, repairs and renovations to them, and often items of new work. The main work section in the Standard Method covering this class of work is section C: Demolition/Alteration/Renovation. Alteration work can be included with demolition work in a single Bill or, alternatively, the alterations and new work can be taken elsewhere with the relevant work sections in accordance with the appropriate rules. For example, forming a door opening in an existing wall involves not only the work of cutting the opening and extending and making good wall and floor finishes, and insertion of a reinforced concrete lintel in accordance with SMM C20.5.2.0.1–3, but also the provision of the door, frame or lining, ironmongery and painting, incorporating items from sections L20 (timber doors), P21 (ironmongery) and M60 (painting).

Blocking up an existing window opening encompasses removal of the window (SMM C20.1.1.0.1), brickwork in filling the opening (SMM C20.8.2.0.1–3), bonding to existing wall (SMM F10.25.1.0.0), wedging and pinning to soffit of opening (SMM F30.7.1.0.0), plastering wall internally (SMM M20.1.1.1.0), and decoration of new plasterwork (SMM M60.1.0.1.0). Decoration of new plaster to filling to openings often involves a specification different from that used for the redecoration of existing painted surfaces.

Example 15 has been designed to incorporate many of the demolition items in SMM C10, alterations in SMM C20, and also a good range of new work, which aims to provide some useful revision practice for the student and covers a number of different forms of construction not measured elsewhere in this book.

The scope and location of the work is to be shown on location drawings or further drawings accompanying the bill of quantities (SMM C10/C20/C40.P1a).

## MEASUREMENT OF DEMOLITION, ALTERATIONS AND REPAIR WORK

### Demolishing structures/shoring

The demolition section often starts with an item giving a general description of the work, which will be supported by the location drawings, which show the location and extent of the existing structures to be demolished (SMM C10.P1a). In addition, any limitations on the method of demolition of the work have also to be stated (SMM C10.S1). For instance, it may be specified that demolition work can only be carried out between the hours of 10.00 a.m. and 4.00 p.m., Mondays to Fridays, or that the use of pneumatic drills is prohibited. Any requirements of this kind must be stated as they will probably result in higher operational costs and the estimator will need to cover them in the billed rates.

The description of demolition work must be sufficient to identify it (SMM C10.1–3.1), and state the levels to which the structures are demolished. Old materials resulting from demolition work frequently become the property of the Contractor, and this is deemed to be the case unless otherwise stated (SMM C10.D1). In like manner, disposal of old materials and temporary support which is at the discretion of the Contractor are deemed to be included in the demolition items (SMM C10.C1).

Demolition item descriptions will include setting aside and storing materials remaining the property of the Employer or those for re-use, such as brickwork for hardcore, the method of demolition where by specific means, or the Employer's restrictions on methods of disposal of materials (SMM C10.S1–3). The handling and disposal of toxic or other dangerous materials, including asbestos, certain chemicals, fuel oils and tars, shall be suitably described, as it may involve the employment of specialist firms or the need for masks or other forms of protective clothing, and it will be necessary to pay attention to the provisions of the Health and Safety at Work Act 1974.

The demolition of structures items should give sufficient details of the buildings or parts thereof involved, although a single item can be provided to cover the work of clearing a site of all buildings. Quantities may be given if considered more appropriate than an item (Code of Procedure: C10). The level(s) to which structures are to be demolished shall be included in the description of the item, and this is often the lowest floor slab. Making good to remaining structures and finishings is also stated in the description (SMM C10.1–3.1.1.1–6).

The support of structures not to be demolished and roads and the like form separate items, stating the type and position of shoring and nature of structure or road, and such other particulars as required (SMM C30.4 –5.1.0.1–6).

*Alterations*
Cutting openings or recesses comprise items and the accompanying dimensioned descriptions shall include the type and thickness of the existing structure, such as one brick wall, and the treatment around the opening or recess, including the insertion of lintels, sills and the like. The supply of these members will, however, generate separate items. Making good structures and extending and making good finishings, such as those to walls and floors, are included in the descriptions (SMM C20.5.2.0.1–3). Students might be tempted to measure these in the finishings section using the normal rules of measurement. Cutting back projections, cutting to reduce the thickness of existing structures and filling in openings are also taken as items (SMM C20.6–8.1–2.0.1–3). Materials arising from alterations (spot items) are the property of the Contractor unless otherwise stated (SMM C20.D1).

Removing fittings and fixtures, including joinery and sanitary appliances, are itemised giving details or a dimensioned description sufficient for identification and, if required, setting aside for re-fixing is included in the description, as in the case of the wood window in example 15. Where it is necessary to make good existing structures or finishings, which can include insulation, this is described (SMM C20.1.1–2.0.1–2). Removing engineering and plumbing installations and removing finishings and coverings to existing structures, such as roofs, are also itemised with similar descriptions (SMM C20.2–4.1–2.0.1–2).

Temporary roofs and screens are itemised with a dimensioned description and including such particulars as required (SMM C20.9–10.1.0.1–6). Alternatively, quantities may be given if considered more appropriate (Code of Procedure). Where temporary arrangements are needed for dealing with rainwater following the removing of roof covering, as in example 15, this is included in the itemised temporary roof item.

*Repairing and renovating concrete, brick, block and stone*
A new section (SMM C40) is incorporated in SMM7 to cover this class of work. Cutting out defective work and replacing with new is suitably described and dimensioned with a choice of measurement units ($m^2$, m and nr) according to the scope of the work (SMM C40.1 & 3.1.1–3.1), and the description includes the method of bonding new to existing (SMM C40.S5 & 6). Resin or cement impregnation/injection items include a dimensioned description, nature of base material, and centres of drilling holes and removing existing finishes (SMM C40.2.1.1–4.1–2). Repointing is measured in $m^2$, stating the size and depth of raking out existing joint, base material and type of pointing, including the composition and mix of mortar (SMM C40.4.1.1–3.1 and C40.S7).

Section C40 also includes rules for the measurement of removing stains (stating the type), cleaning surfaces (stating the method), inserting new wall ties (stating size and type of tie, nature of walling and surface finishes), redressing to new profile, and artificial weathering (SMM C40.5–9).

Section C41 covers the provision of chemical damp-proof courses, measured in metres, stating the method of operation (where by specific means), chemicals, nature and thickness of wall, centres of drilling holes and removing existing finishes (SMM C41.1.1–3.1.1–2). Sections C50, 51 and 52 cover the repairing or renovating of metal or timber and the treatment of existing timber, such as the eradication of fungus attack or beetle infestation.

## SEQUENCE OF MEASUREMENT

No hard and fast rules of procedure can be prescribed for the measurement of this class of work. A systematic and logical approach is essential and the preparation of a take off list, as incorporated in example 15, is a great help and this should be supported by the crossing-off of items on the drawings as they are taken off. All possible steps should be taken to endeavour to eliminate the omission of items.

The sequence of taking off adopted in example 15 provides a good approach to this class of work. It starts with demolition items and filling openings and then follows with new work in the order adopted in SMM7. The student must take care to include the work around the periphery of new openings and filling of old openings.

The removal of redundant rainwater goods and gullies and the sealing-off of old drains which are no longer required must be included, while the removal of coverings, finishings, fixtures and fittings all find a place in the measurement of this type of work.

The new work follows a logical sequence and the dimensions, descriptions and explanatory notes in example 15 should provide useful revision of some of the more important work sections. Furthermore, some additional forms of construction are covered such as cavity fill, bonding new brick and concrete block skins to existing brick walls, isolated pier, granolithic paving, bitumen felt flat roof, composition steel roof decking, drain connections and air test to drains.

The student is advised to read carefully through example 15, referring to the drawing as appropriate.

## REVISION NOTES

Some of the more crucial weaknesses identified in examiners' reports for the subject of Building Measurement in the final examinations of the principal professional bodies are included as they highlight problem areas and will cause students to reflect on their own approach, avoid common pitfalls and assist with the work of revision.

## GENERAL ASPECTS

*Need to adopt good practice*
Many candidates failed to realise the importance of adopting recognised good practice when taking off dimensions. It is important, as shown throughout the worked examples, to number pages, bracket where appropriate, annotate, carry out waste calculations and adopt a logical order of approach.

Timesing of dimensions was often incorrect and some items that should have been measured in $m^3$ were frequently added to items measured in $m^2$. It is good policy to check all measured items to ensure that the correct unit of measurement is being used. When a question reads 'including all necessary adjustments', deductions for window and door openings and all associated labours must be measured. Some poor results stemmed from a lack of understanding of the Standard Method and from an inability to write adequate descriptions and to read drawings correctly.

*Accuracy of dimensions*
Few candidates consistently produced correct calculations and accuracy is vitally important as inaccurate quantities are worthless. It is not good enough for candidates to adopt such slapdash methods as to deduct 25 per cent for overmeasure or to add 25 per cent for undermeasure. On the other hand the student should avoid spending a long time to obtain minute dimensions. Avoid splitting hairs, adopt a realistic approach and keep a sense of proportion. Where a candidate runs out of time before completing the paper, it is wise to enter the remaining items in the form of a take off list.

*Presentation and layout*
There appears to be considerable room for improvement in the presentation of measured work, particularly in the detailing of side notes, analysis of sections, use of headings and sub-headings and general neatness. One examiner observed: "If some candidates had to return to their papers after the completion of the contract, for the computation of the final account, they would be at a loss to reconcile what and where they had measured, and candidates must realise that they have a better chance of passing if the examiner can read the script."

The standard of presentation was generally considered to be poor, with a lack of waste calculations and signposting of dimensions. In some instances the drawings were seriously misread and few candidates made any attempt to raise queries with the architect, even when requested to do so in the rubric to the examination paper.

*Preparation of schedules*
When preparing schedules avoid unnecessary duplication of information. For instance, to write the same description for finishings for each identical item, when they could have been collated, is a waste of valuable examination time.

*Examination techniques*
It is good practice to read the questions and drawings carefully to identify clearly what is to be measured, to prepare draft notes or take off lists, and to allocate the examination time over the questions in proportion to their mark allocations. Some lecturers recommend reading an examination question once again when it is half answered. At this stage a candidate who has been working at the question is more receptive to subtle meanings and to the

specific requirements of the examiner than when he first read the question. If he finds that he has misinterpreted part of the question, he may still have time to correct his error(s). It is especially important in a measurement paper to read every word on the drawing(s) carefully and deliberately.

*Systematic and logical approach*
Students must recognise that measurement is essentially a communication process, requiring the presentation of information and data in a coherent and disciplined way. The process should accordingly be as efficient as possible and the results accurate. In this respect it helps considerably if the layout and presentation are clear and legible, assisting, and certainly not hindering, the easy assimilation and retrieval of information in the take off. One examiner emphasised yet again the importance of legibility and the need for the student to cultivate it. He also referred to the inability of many examination candidates to organise their work in a satisfactory way. Most examination syllabuses stress the importance of a 'systematic and logical approach to the analysis of project data', but many candidates fail to do this. Disorganised and untidy work loses vital marks in the examination.

Every examiner's report refers to poor presentation. One examiner stated that he did not expect scripts of copperplate quality, but they must be readable. It was also emphasised that a generous spacing not only aids clarity, but also allows some latitude for the correction of errors. Greater use of waste calculations to support dimensions and increased 'signposting' helps the examiner to follow the candidate's approach and train of thought.

The actual dimensions may be taken from those figured on the drawings, but frequently some dimensions require a basic calculation in waste — the most common example being a centre line or girth, and these must always be clearly shown. Many candidates seem reluctant to include side calculations and, where they are provided, they are often relegated to an obscure corner.

*Knowledge of the Standard Method*
The number of deviations from the Standard Method was very extensive and students are strongly recommended to spend more time reading and digesting the contents of the document, its requirements and application to measurement practice. Far too much work submitted by examination candidates ignores or contradicts requirements of the Standard Method. Students would also benefit from a close study of the *SMM7 Code of Procedure for Measurement of Building Works*.

*Examination approach*
Candidates will benefit from the adoption of a methodical approach to measurement questions in the examination. The following suggested approach may prove helpful to the candidate.

(1) Identify the work to be measured by a careful examination of the question, drawings and other supporting information and formulate a plan of operation. This is often in the form of a take off list and two are incorporated in examples 7 and 15. The constraints imposed by the Standard Method need to be borne in mind in its preparation. This provides the basis for an orderly approach and a useful checklist, prior to becoming engrossed at the detailed level of dimensions and descriptions. It also provides a positive safeguard against the omission of major items.

(2) Quantify and set down clearly the relevant dimensions and frame suitable descriptions in accordance with the prescribed rules. Full use should be made of sub-headings and side wastes to clarify each step in the measurement process. The detailed Standard Method requirements, such as material descriptions and depth and thickness classifications, must be implemented.

*Query sheets*
The examiners endeavour to provide good working drawings which are complete in every detail and which are capable of only one interpretation. As in practice, errors, omissions and misunderstandings are almost certain to occur. Candidates should seek to clarify doubtful points in accordance with normal office practice, and should draft a query sheet, ask the questions and answer them. This allows the examiner to understand the reasons for a particular item or method of measurement. The query lists should be regarded as an essential part of the measurement process.

## WEAKNESSES IN SPECIFIC WORK SECTIONS

*Excavating and filling*
The measurement rules for excavating and earthwork support, and working space requirements generally, need to be well known and fully understood. Some candidates measured items of trench excavation, backfill and disposal of excavated material to exactly the same dimensions, and

a frequent mistake was to use the same height for measuring brickwork up to damp-proof course level as for backfill. Some candidates made no adjustment for concrete foundations in the disposal of excavated material.

In many instances no maximum depth classification of trenches or earthwork support was given. Unrealistic specification assumptions were made by some candidates and time was wasted in measuring excavations below groundwater level and in unstable ground where they did not apply. One candidate, however, decided to save time by assuming that all excavation (the subject of the question) had been measured previously!

It is worth stressing that topsoil excavation is measured only if it is to be preserved and even then only if it exists. When measuring basement excavation few candidates considered the possibility of measuring reduce level excavation to level the site before measuring basement excavation. Where this was done it considerably simplified the remainder of the excavation measurement. Some candidates continue to measure working space excavation as three items — excavation, backfill and disposal — instead of one working space allowance to excavations item. Descriptions that are inappropriate are also included on occasion, such as 'compacting the base of a small pit with a 2.5 tonne roller'.

The measurement of reduce level excavation for a site with a grid of nine levels presented a major problem for many candidates. The examiner suggested the following logical approach.

(1) Establish the required formation level.
(2) Check whether any grid level or levels will be below this formation level after removal of topsoil.
(3) If there are levels below formation there will be filling and these levels should not be used to establish the average level for excavation.

In this particular question, one level was below formation and it was necessary to establish an approximate cut and fill line. A simple average of the grid levels was adequate as the graduation between levels was relatively uniform, although a weighted average provided a better solution.

*Concrete work*
Descriptions were frequently poor and some candidates measured concrete beds as foundations or ground beams and others measured them in m$^2$ with thickenings measured separately. Surface labours to have any meaning to an estimator cannot be included in the description of concrete work measured in m$^3$. When formulating descriptions candidates must include all requirements of the Standard Method, such as thickness categories of concrete. Some failed to classify concrete and formwork in the categories listed in the Standard Method.

It is not satisfactory to measure bar reinforcement simply by inserting a note to the worker up to abstract the information for the reinforcement from the bending or bar schedule, nor is it sufficient to copy the reinforcement from the schedule without the inclusion of proper descriptions. It is not necessary to state how many times each bar is bent. Where lengths of bars are extracted from bending schedules, no additional allowance is required for hooks and bends. Where no schedules were provided, it was surprising how few candidates acknowledged the fact that steel reinforcement requires a cover of concrete at the edges of members.

Formwork is measured as the actual surfaces of the finished structure which need to be supported when depositing the concrete.

*Brickwork*
Brick walls in facings are measured in accordance with SMM F10.1.2–3.1.0, and facework is no longer measured as extra over common brickwork. It is necessary to be able to identify isolated piers and to deal with projections and facework ornamental bands in a satisfactory manner.

*Other work sections*
Candidates should note the special *asphalt* requirements in SMM J20/J21.P1a to provide a plan of each level and its height above ground level together with any restrictions on the siting of plant and materials and a section indicating the extent of tanking work. With *slate* or *tile roofing*, the particulars of the roof covering must include the minimum laps (SMM H60/H61.S3), and work to different pitches is given separately (SMM H60/H61.1.1.0.0).

With *carpentry* timbers it is only necessary to distinguish between roof timbers in flat and pitched roofs, and not individual members except plates. Students should note the requirement to separate timbers in continuous lengths > 6.00 m (SMM G20.6–10.1–2.1.1), and that work is deemed to include labours subject to a few exceptions (SMM G20.C1).

The application of *composite items* for composite work manufactured off site in accordance with SMM General Rules 9.1 should be clearly understood, and this procedure is illustrated in example 11.

*Structural metalwork* is classified under structural steel/ aluminium framing (fabrication and erection) and isolated structural steel members (SMM G10, 11 and 12). Candidates sometimes fail to include fittings after the member(s) with which they are associated. It should also be remembered that certain items are enumerated and described separately, such as fixing bolts for isolated structural members (SMM G12.D8 and M7) and purlins and cladding rails in metres (SMM G10.4.1.0.0).

Some candidates experienced difficulty with a basic *electrical installation* question. The major faults were as follows:

(1) Failure to identify work under the appropriate headings.

(2) A lack of understanding of SMM Y61.19.1–2.1 –6.1–5, which requires final circuits forming part of a domestic or similar simple installation from distribution boards and the like to be enumerated stating the number of points.

(3) Ignoring associated builder's work.

(4) A lack of order and method in the measurement resulting in omission of work.

On occasions candidates omitted to classify external work separately in *finishings*, *painting* and *decorations*. Full particulars of each type of finishing and decorative treatment must be given in accordance with the requirements of the Standard Method. For instance, in one examination quarry tiles were described as 'clay quarry tiles', without mention of quality, size, thickness, nature of base, bedding, and treatment and layout of joints as SMM M40.S1–8. Care must be taken to separate and describe work in staircase areas and plant rooms. Some candidates fail to separate beds and backings (screeds) from finishings and are unable to determine the type of finish required to screeds.

*Dimensioned and bill diagrams*
In all cases bill items must be adequately described to comply with SMM7, but in certain cases, such as doors and windows, the Standard Method calls for a dimensioned diagram where an adequate description could prove difficult or lengthy and where drawn information is likely to give a clearer understanding of the item(s). In many of these cases, reference to an architect's drawing will be sufficient and copies of these drawings should be sent to tenderers with the bill of quantities. For minor details, however, such as labours to natural stone, a dimensioned diagram as described in SMM General Rules 5.3, taken from the relevant drawing, will often be more suitable. Examples of dimensioned and bill diagrams appear throughout the worked examples in this book.

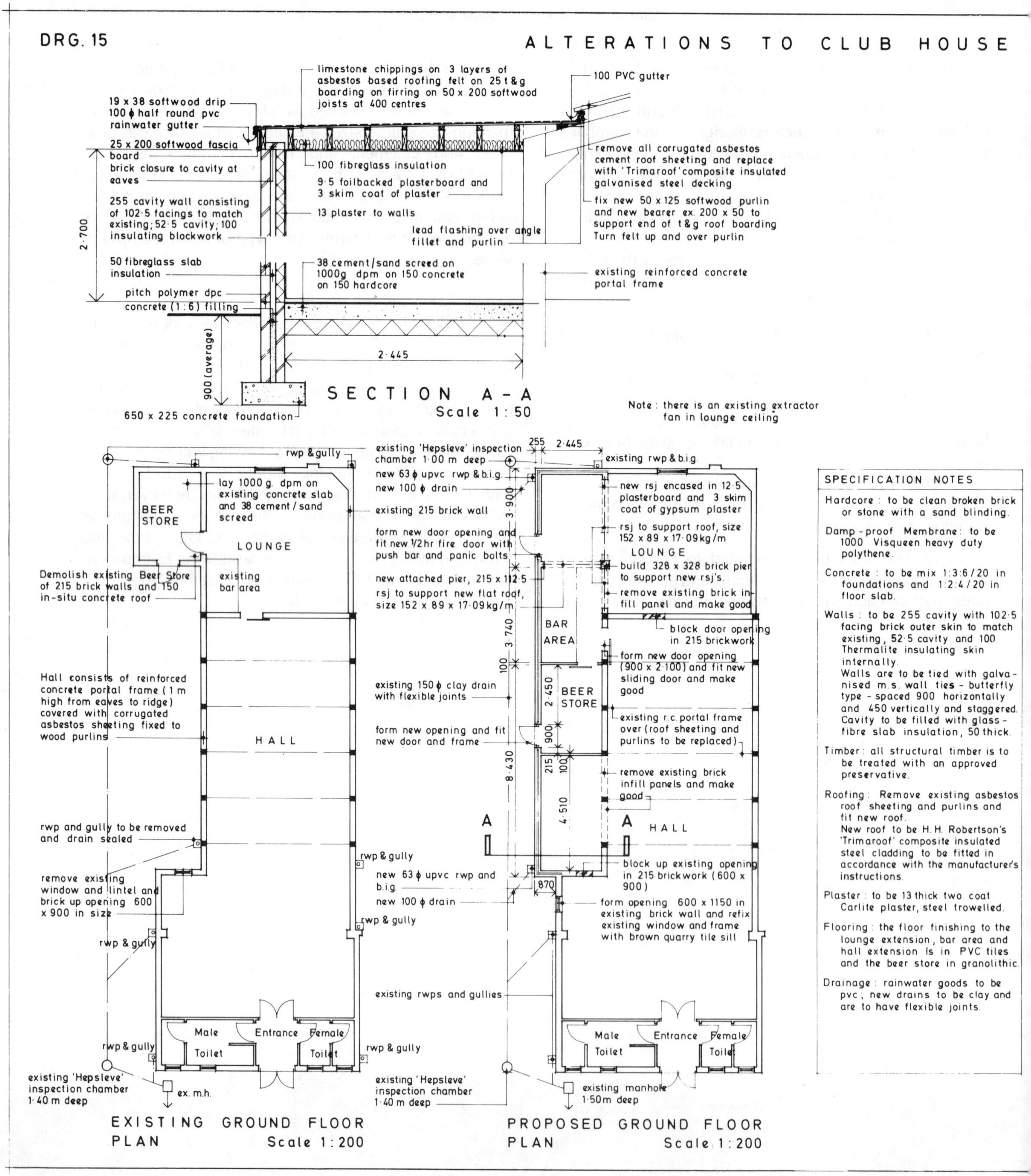
DRG. 15
ALTERATIONS TO CLUB HOUSE
limestone chippings on 3 layers of asbestos based roofing felt on 25 t & g boarding on firring on 50 x 200 softwood joists at 400 centres
100 PVC gutter
19 x 38 softwood drip
100 ϕ half round pvc rainwater gutter
25 x 200 softwood fascia board
brick closure to cavity at eaves
100 fibreglass insulation
9·5 foilbacked plasterboard and 3 skim coat of plaster
remove all corrugated asbestos cement roof sheeting and replace with 'Trimaroof' composite insulated galvanised steel decking
255 cavity wall consisting of 102·5 facings to match existing; 52·5 cavity; 100 insulating blockwork
13 plaster to walls
fix new 50 x 125 softwood purlin and new bearer ex. 200 x 50 to support end of t & g roof boarding
Turn felt up and over purlin
lead flashing over angle fillet and purlin
2·700
50 fibreglass slab insulation
38 cement/sand screed on 1000g dpm on 150 concrete on 150 hardcore
existing reinforced concrete portal frame
pitch polymer dpc
concrete (1:6) filling
900 (average)
2·445
650 x 225 concrete foundation
SECTION A-A
Scale 1:50
Note: there is an existing extractor fan in lounge ceiling
rwp & gully
existing 'Hepsleve' inspection chamber 1·00 m deep
new 63 ϕ upvc rwp & b.i.g.
new 100 ϕ drain
existing 215 brick wall
lay 1000 g. dpm on existing concrete slab and 38 cement/sand screed
BEER STORE
LOUNGE
form new door opening and fit new 1/2 hr fire door with push bar and panic bolts
Demolish existing Beer Store of 215 brick walls and 150 in-situ concrete roof
existing bar area
new attached pier, 215 x 112·5
rsj to support new flat roof, size 152 x 89 x 17·09 kg/m
255
2·445
existing rwp & b.i.g.
new rsj encased in 12·5 plasterboard and 3 skim coat of gypsum plaster
rsj to support roof, size 152 x 89 x 17·09 kg/m
LOUNGE
build 328 x 328 brick pier to support new rsj's.
remove existing brick infill panel and make good
3·900
3·740
100
BAR AREA
block door opening in 215 brickwork
form new door opening (900 x 2·100) and fit new sliding door and make good
Hall consists of reinforced concrete portal frame (1 m high from eaves to ridge) covered with corrugated asbestos sheeting fixed to wood purlins
existing 150 ϕ clay drain with flexible joints
2·450
BEER STORE
existing r.c. portal frame over (roof sheeting and purlins to be replaced)
form new opening and fit new door and frame
900
HALL
8·430
215
100
remove existing brick infill panels and make good
4·510
A
A
HALL
rwp and gully to be removed and drain sealed
rwp & gully
new 63 ϕ upvc rwp and b.i.g.
block up existing opening in 215 brickwork (600 x 900)
870
remove existing window and lintel and brick up opening 600 x 900 in size
new 100 ϕ drain
form opening 600 x 1150 in existing brick wall and refix existing window and frame with brown quarry tile sill
rwp & gully
rwp & gully
existing rwps and gullies
Male Toilet
Entrance
Female Toilet
rwp & gully
rwp & gully
Male Toilet
Entrance
Female Toilet
existing 'Hepsleve' inspection chamber 1·40 m deep
ex. m.h.
existing 'Hepsleve' inspection chamber 1·40 m deep
existing manhole 1·50m deep
EXISTING GROUND FLOOR PLAN Scale 1:200
PROPOSED GROUND FLOOR PLAN Scale 1:200
SPECIFICATION NOTES
Hardcore: to be clean broken brick or stone with a sand blinding.
Damp-proof Membrane: to be 1000 Visqueen heavy duty polythene.
Concrete: to be mix 1:3:6/20 in foundations and 1:2:4/20 in floor slab.
Walls: to be 255 cavity with 102·5 facing brick outer skin to match existing, 52·5 cavity and 100 Thermalite insulating skin internally.
Walls are to be tied with galvanised m.s. wall ties - butterfly type - spaced 900 horizontally and 450 vertically and staggered. Cavity to be filled with glass-fibre slab insulation, 50 thick.
Timber: all structural timber is to be treated with an approved preservative.
Roofing: Remove existing asbestos roof sheeting and purlins and fit new roof.
New roof to be H. H. Robertson's 'Trimaroof' composite insulated steel cladding to be fitted in accordance with the manufacturer's instructions.
Plaster: to be 13 thick two coat Carlite plaster, steel trowelled.
Flooring: the floor finishing to the lounge extension, bar area and hall extension is in PVC tiles and the beer store in granolithic.
Drainage: rainwater goods to be pvc; new drains to be clay and are to have flexible joints.

## ALTERATIONS

The wk. comprises demolitn. alteratns. & new wk. to the Club Hse. to provide extns. to the xtg. bldg., as shown on drg. 15. Demolitn. wk. shall be carried out between the hrs. of 9.00 am and 5.00 pm. Mon.–Fri. & pnematic drills are not to be used.

A general description of the work has been given, including any limitations on the method of working, to assist the estimator in pricing the work, as provided for in SMM A13. Old materials arising from demolition work and alterations are the property of the Contractor unless otherwise stated (SMM C10/C20. D1). The arrangement for the provision of credits in the bill for materials which are the property of the Contractor no longer applies. Demolition work and alterations have been separated from new work.

### Demolition & alteratns.

All mats. resultg. from the demolitn. & alteratns. which are not specified as being re-used or to remain the property of the Employer, are the property of the Contractor and shall be removed from the site.

As the work is rather fragmented it would be good practice to prepare a take off list, to reduce the risk of omission of items. A suitable list follows:

### Demolition & alteratns.

Demolish beer store
Remove bk. infill pans.
Block up dr. opg.
Remove wdw. & blk. up opg.
Form new wdw. opg., etc.
Remove d.p. & gully & seal dm.
Remove purlins
Remove asb. rf. coverg.

### New wk.

Topsoil removal
Fdns.
Ext. wall
Pier
Int. walls
Floor slab
Stlwk.
Flat rf. & r.w. drainage
Pitched rf. coverg. etc.
Doors
Finishgs.
Drainage

### Demolition & alteratns.

| | | |
|---|---|---|
| Item | Demolish xtg. single storey beer store 3.9 m × 2.9 m × 2.5 m hi. down to the top of the conc. flr. slab & m/gd. structure disturbed. All bwk. is permitted to be used as h.c. in the new wk. | Demolition of individual structures, or parts thereof, are given as items, including a description sufficient for identification, and the level to which they are to be demolished. Any remaining structures to be made good and re-use of materials are included in the description (SMM C10.2.1.1. 2–3). Old materials to remain the property of the Employer are so described (SMM C10. D1). |

15.1

## EXAMPLE 15

| | | |
|---|---|---|
| Item | Cuttg. out 2 nr 1B infill pans. 1.65 m × 2.60 m. hi., between xtg. r.c. portal frs., m/gd. struct. & extend & m/gd. flr. fin. of PVC tiles. | Cutting openings or recesses in existing structures are given as items, with a dimensioned description, including the type and thickness of the existing structure, making good the structure and extending and making good finishings around the opening or recess (SMM C20.5.2.0.1–2). Inserting new work can be included in the description. |
| Item | Cuttg. out len. of 1B wall, 1.65 m × 2.60 m hi., inc. formg. & pla. to one sq. jamb & extend & m/gd. flr. fin. of PVC tiles. | |
| | add lintel  2.100  150  ─────  2.250 | |
| Item | Cuttg. out dr. opg. in 1B wall, 900 × 2.250 inc. formg. & pla. to 2 sq. jambs & insertg. r.c. lintel 215 × 150 × 1200 lg. (supply m/s) & extend & m/gd. flr. fin. of PVC tiles. | The item for cutting out the door opening includes the treatment of reveals, the insertion of the lintel and extending and making good floor finishings (SMM C20.5. 2.0.1–3). The height of the door opening has been increased to include the depth of the lintel. |
| Item | Fillg. in opg., 900 × 1200 w. 1 B wall in comms. in c.m. (1:4). | Filling in openings is given as an item with a dimensioned description as SMM C20.8.2. 0.0. |
| 2/ 2.10 | Bonding end of new 1B wall to xtg. bwk. | Bonding brickwork at sides of opening as SMM F10.25. 1.0.0. Forming chases and holes are deemed to be included (SMM F10.C1c). |
| 0.90 0.22 | Prepg. top of xtg. 1B wall to rec. new bwk. | Preparing tops of existing walls for raising is measured as SMM F10.26.1.1.0. |
| 0.90 | Wedgg. & pinng. new 1B wall to u/s of xtg. wall in slates in c.m. (1:2). | Wedging and pinning new work to the underside of existing construction constitutes a separate linear item (SMM F30.7.1.0.0). |

15.2

## ALTERATIONS (Contd.)

| | | | |
|---|---|---|---|
| 2/ | 0.90<br>2.10 | Pla. to walls, width > 300, 2 cts. Carlite 13th. %a to bwk. in fillg. to opg. trowlld. fin. | Plaster to walls is measured in accordance with SMM M20.1.1.1.0, including nature of base as SMM M20.S5. |

$$\begin{array}{r} 900 \\ \text{add lintel} \quad \underline{150} \\ \underline{1.050} \end{array}$$

| | | | |
|---|---|---|---|
| | Item | Cuttg. wdw. opg. in 1B wall, 600 × 1050, inc. formg. sq. jambs & extendg. & m/gd. finishgs. & insertg. r.c. lintel, 215 × 150 × 900 lg. (supply m/s). | The work is deemed to include fair joints (SMM M20.C1a).<br><br>The item for forming the window opening, includes extending and making good finishings and inserting a reinforced concrete lintel (SMM C20.5.2.0.1-3). |
| | Item | Removg. xtg. wd. cast. & fr., 600 × 900 %a & set aside for re-fixg. | Item for removing existing window and setting aside for re-use as SMM C20.1.2.0.0 and C20.S2. |
| | 1 | Refix wd. cast. & fr. 600 × 900 %a, in prepd. opg. in xtg. bk. wall, inc. beddg. & ptg. fr. | Enumerated item for refixing existing wood casement and frame. |
| | 1 | Supply only precast conc. lintel 215 × 150 × 1200 lg. (1:1½:3/20 agg.) reinfd. w. 2 nr 12⌀ m.s. bars to BS 4449, w. 3 nr surfs. (515%a gth.) keyed for pla. (dr. opg. | The supply of lintels to the door and window openings are taken as enumerated items, with a dimensioned description, including reinforcement details as SMM F31.1.1.0.1. Formwork is deemed to be included in the items (SMM F31.C1). |
| | 1 | Ditto. 215 × 150 × 900 lg. reinfd. a.b. w. 2 nr. surfs, (365 %a gth.) keyed for pla. & 1 nr. surf. (150 %a gth.) fin. smth.   (wdw. opg. | The lintel to the window opening is varied as necessary. It seems desirable to complete the window items at this stage to avoid having to return to them later. |
| | 0.60 | Sill brown quarry tiles, width: 120, 20th., b. & p. in c.m. (1:3) w. flush jts. on screeded bed (m/s)<br>&<br>Screed to margin, width: 120, 10 ct. & sd. (1:2) in 1 ct. to rec. quarry tiles ld. lev. on bwk. | The sill is measured in metres, stating the width as SMM M40.7.0.1.0, giving the particulars listed in SMM M40.S1-7. Cutting at ends is deemed to be included (SMM M40.C1d).<br><br>The bed is also measured in metres and classified as SMM M10.11.0.1.0 (margins) |

15.3

| | | | |
|---|---|---|---|
| | 0.60<br>0.90 | Bk. wall, facewk. o.s., to match xtg., thickness: 215 in Flem. bond in g.m. (1:1:6) in fillg. to xtg. opg. & ptg. w. a flush jt.<br>&<br>Pla. to walls, width > 300, 2 cts. Carlite 13th. %a to bwk. in fillg. to opg., w. trowlld. fin. | as the nearest appropriate item and giving the particulars listed in SMM M10.S1-6.<br>The brick filling to the existing window opening is measured in m² stating the thickness and classified as SMM F10.1.2.1.0 and giving the appropriate particulars listed in SMM F10.S1-5.<br>Plasterwork suitably described as before. |
| 2/ | 0.90 | Bondg. end of new 1B wall to xtg. bwk. | Bonding brickwork at sides of opening as SMM F10.25.1.0.0. |
| | 0.60<br>0.22 | Prepg. top of xtg. 1B wall to rec. new bwk. | Preparing top of existing wall for raising as SMM F10.26.1.1.0. |
| | 0.60 | Wedgg. & pinng. new 1B wall to u/s of xtg. wall in slates in c.m. (1:2). | Wedging and pinning new work to underside of existing construction as SMM F30.7.1.0.0. |
| | Item | Removg. 2.90 m len. of xtg. PVC downpipe & m/gd. bwk. | Removing plumbing installations are given as items with sufficient details for identification (SMM C20.2.1.0.1). |
| | Item | Removg. xtg. b.i. gully & seal off xtg. drain. | Making good existing structure or finishings are included in the description. |
| | Item | Removg. xtg. bar fittgs. 2.15 m × 1.30 m in area & m/gd. xtg. wall & flr. finishgs. The fittings are to remain the property of the Employer. | The removal of fittings and fixtures are given as an item, with a dimensioned description. The description is to include any making good to existing structure or finishings (SMM C20.1.2.0.1-2). Old materials to remain the property of the Employer are so described (SMM C20.D1 and S2). The length of each purlin is calculated in waste from the figured dimensions on the drawing. |

$$\begin{array}{r} \text{Roofg. to Hall} \\ \text{purlins} \\ 8.430 \\ 100 \\ 3.740 \\ \underline{3.900} \\ 16.170 \\ \text{less} \\ \text{outer bk. skin} \quad \underline{100} \\ 8/\underline{16.070} \\ \text{total len.} \quad \underline{128.560} \end{array}$$

15.4

## ALTERATIONS (Contd.)

| | | |
|---|---|---|
| Item | Removg. xtg. 50 × 125 swd. purlins fxd. to r.c. portal fr. (129 m total len.) & m/gd. xtg. structure. | The removal of the purlins is given as an item in accordance with SMM C20.1.1.0.1. The timber will be deemed to become the property of the Contractor, unless otherwise stated, in accordance with SMM C20.D1. |

$$\begin{array}{r}\text{Asb. sheetg.}\\\underline{\text{len.}}\\16\cdot170\\\text{add o/hg. 2/75}\quad\underline{\;\;\;150\;\;\;}\\16\cdot320\\\underline{\text{Area}}\\2/\overline{16\cdot320}\\\underline{\;\;3\cdot500\;\;}\\114\cdot240\end{array}$$

The area of the existing asbestos cement roof covering is calculated in waste.

| | | |
|---|---|---|
| Item | Removg. xtg. corrugated asb. ct. rf. coverg. (115 m² total area). | Removing coverings to existing structures shall be given as items, giving a dimensioned description as SMM C20.4.2.0.0. |
| Item | Tempy. roof (115 m² total area) providg. & erectg., maintaing. as nec. & clearg. away on completn. & disposg. of rainwater. | Temporary roof item for dealing with rainwater as SMM C20.9.1.0.1–5. |

$$\begin{array}{r}\text{New wk.}\\\underline{\text{Topsoil}}\\\underline{\text{len.}}\\3\cdot740\\100\\8\cdot430\\\underline{\;\;\;\;\;\;\;\;}\\12\cdot270\\\underline{\text{width}}\\\text{add. fdn. sprd.}\quad 2\cdot445\\650\\\underline{\;\;\;255\;\;\;}\\2/\overline{395}\quad\underline{\;\;\;198\;\;\;}\\2\cdot643\end{array}$$

The measurement of the new work will now be taken in a logical sequence and will provide some useful revision practice for the student, as it embraces a good range of work sections. The small adjustment for the short return wall is not considered worth making, especially as it is already largely offset by the length that is being taken. The student must always keep a sense of proportion.

| | | |
|---|---|---|
| 12·27<br>2·64 | Exc. topsoil for preservn. av. 150 dp. | Excavating topsoil to be preserved is measured in m² stating the average depth (SMM D20.2.1.1.0). |
| 12·27<br>2·64<br>0·15 | Disposal of excvtd. mat. on site in spoil hps. av. dist. of 20 m from excavn. | Disposal measured in accordance with SMM D20.8.3.2.1. Some will be needed to subsequently backfill the foundation trench. |

15·5

---

$$\begin{array}{r}\underline{\text{Fdn.}}\\\underline{\text{len.}}\\12\cdot270\\\underline{\;\;\;870\;\;\;}\\13\cdot140\\\text{less corner}\quad\underline{\;\;\;255\;\;\;}\\12\cdot885\\\underline{\text{depth}}\\900\\\text{less topsoil}\quad\underline{\;\;\;150\;\;\;}\\750\end{array}$$

The length of foundation trench is not extended around the outside of the old beer store, as it is assumed that the new wall will be built off the base of the old wall.

| | | |
|---|---|---|
| 12·89<br>0·65<br>0·75 | Exc. tr. width > 0·30 m., max. depth ≤ 1·00 m.<br>&<br>Fillg. excavns., av. thickness > 0·25 m, arisg. from excavns. | Excavation to trenches is measured in the prescribed stages of maximum depth and stating the width classification as SMM D20.2.6.2.0. All excavated soil is taken as filling returned to trenches in the first instance (See SMM D20.9.2.1.0). The origin of the filling material shall be one of the three categories in SMM D20.9.2.1–3. |
| 12·89<br>0·65 | Compactg. bott. of excavn. | Compacting of bottoms of excavation is measured in m² in three separate categories (SMM D20.13.2.1–3.0). Earthwork support is measured in m² stating the maximum depth range and the distance between opposing faces (SMM D20.7.1.1.0). |

$$\begin{array}{r}12\cdot885\\\text{add}\quad\underline{\;\;\;325\;\;\;}\\\text{outer face of tr.}\quad 13\cdot210\\12\cdot885\\\text{less}\quad\underline{\;\;\;325\;\;\;}\\\text{inner face of tr.}\quad 12\cdot560\end{array}$$

| | | |
|---|---|---|
| 13·21<br>1·00<br>12·56<br>0·75 | Earthwk. suppt. max. depth ≤ 1·00 m / outer dist. between (face opposg. faces ≤ 2·00 m. (inner (face | The earthwork support to the outer face of the trench will include the depth occupied by topsoil. This can be strutted from the opposing face of the trench and so the distance of ≤ 2·00 m can apply. |
| 12·89<br>0·65<br>0·23 | In situ conc. fdns. (1:3:6/ 20 agg.), poured on or against earth.<br>&<br>Ddt. Fillg. to excavns. a.b.<br>&<br>Add Disposal of excvtd. mat. off site. | Note the classification of in situ concrete in foundations as SMM E10.1.0.0.5. The concrete is followed by adjustment of soil disposal. The rough finish to foundation concrete to receive walling is outside the scope of surface treatments in SMM E41.1–7. |

15·6

ALTERATIONS (Contd.)

| | | | |
|---|---|---|---|
| | | bwk ht. | |
| | | 900 | |
| | less fdn. | 225 | |
| | | 675 | |
| | add above g.l. | 150 | |
| | | 825 | |

2/ 12·89
0·83
Bk. wall, thickness: 102·5, in comms. in stret. bond in c.m. (1:3).

Height of brickwork is calculated in waste. Half brick skins of hollow walls are measured as half brick walls in m² (SMM F10.1.1.1.0 and F10. D4). Alternatively, the wall could be described as ½B thick. The part of the outer brick skin built in facework will be adjusted later. Brick dimensions are given in a preamble clause and are normally 215 × 102·5 × 65. The description of brickwork includes bricks, bond and mortar, and pointing where appropriate (SMM F10.S1-4). The formation of the cavity, stating the width, type, size and spacing of wall ties are given in a single superficial item (SMM F30.1.1.1.0). Note the reference to the British Standard in the description of the wall ties. Alternatively, there could be a reference to the project specification or a preamble clause and use can be made of the SMM 7 Library of Standard Descriptions. Concrete filling to cavities of hollow walls is measured in m³ with the thickness range given as SMM E10.8.1-3.

12·89
0·83
Form cav. in holl. wall width: 50 inc. galv. m.s. butterfly wall ties to BS 1243, spaced 900 hor. & 450 vert. & staggd.

12·89
0·05
0·68
In situ conc. (1:6) fillg. holl wall, thickness ≤ 150.

| | | |
|---|---|---|
| | | 675 |
| less topsoil | | 150 |
| | | 525 |

12·89
0·26
0·53
Ddt. Fillg. to excavns. a.b.
&
Add Disposal of excvtd. mat. off site.

Adjustment of excavated soil disposal for volume displaced by brickwork below topsoil.

2/2/ 0·83
Bondg. end of ½B wall to xtg. bwk.

Bonding ends of new walls to existing are given in metres stating the thickness of the new work (SMM F10. 25.1.0.0). Forming holes for pockets are deemed to be included (SMM F10. C1c).

15. 7

---

2/ 12·89
0·10
D.p.c. width ≤ 225, hor., pitch polymer bedded in c.m. (1:3).

| | |
|---|---|
| | 3·900 |
| | 2·445 |
| | 6·345 |

6·35
0·22
(former beer store)

| | facewk. ht. |
|---|---|
| above grd. | 150 |
| below grd. | 75 |
| | 225 |

12·89
0·23
Ddt. Bk. wall, thickness: 102·5, comms. in c.m. (1:3).
&
Add Ditto. facewk. o.s., in fcg. bks. to match xtg. in stret. bond in c.m. (1:3) & ptg. w. a nt. flush jt. as wk. proceeds.

| | |
|---|---|
| | 13·210 |
| less fdn. sprd. | 198 |
| | 13·012 |

13·01
0·20
0·15
Fillg. to excavns., av. thickness ≤ 0·25 m from on site spoil hps., topsoil.

Item    Disposal of surf. water.

| | Ext. wall above dpc | |
|---|---|---|
| | | 12·885 |
| | | 6·345 |
| mean gth. | | 19·230 |
| ½ th. of cav. | | 26·25 |
| ½ th. of h.b. skin | | 51·25 |
| | | 77·50 |
| ½ th. of cav. | | 26·25 |
| ½ th. of blkwk. | | 50·00 |
| | | 76·25 |

Damp-proof courses ≤ or > 225 wide are each measured separately in m² giving the particulars listed in SMM F30. 2.1-2.1-4.1, as appropriate. Pointing exposed edges is deemed to be included and no allowance is made for laps (SMM F30.C2 and M2). It is usual to allow one course of facings below ground level to counteract any irregularities in the finished ground level. The facework previously measured as common brickwork is now adjusted. The centre line girth has not been adjusted because of the small quantity involved. Always keep a sense of proportion. This item includes the facing bricks which are not measured as extra over common brickwork. Note the method of arriving at the mean girth of foundation spread outside the building, working from the outside trench face previously calculated for earthwork support. Adjustment of topsoil outside the building, obtaining the topsoil from on site spoil heaps (SMM D20.9.1.2.3). See SMM D20.8.1.0.0 for surface water disposal. It is assumed that the excavation work is above normal groundwater level, otherwise an additional item would be required in accordance with SMM D20.8.2.0.0. Calculation of girths of cavity, outer brick skin and inner skin of concrete blocks in waste.

15. 8

ALTERATIONS (Contd.)

|  |  |  |
|---|---|---|
|  | outer skin |  |
| add | 19.230 |  |
| 2 corners 2/2/77.50 | 310 |  |
|  | 19.540 |  |
|  | inner skin |  |
| less | 19.230 |  |
| 2 corners 2/2/76.25 | 305 |  |
|  | 18.925 |  |

| | | |
|---|---|---|
| 19.54 / 2.70 | Bk. wall, thickness: 102.5, facewk. o.s., in fcg. bks. to match xtg. in stret. bond in g.m. (1:1:6) & ptg. w. a nt. flush jt. as wk. proceeds. | Bricks will normally be described and often accompanied by a basic price or prime cost per 1000 bricks delivered to the site, to assist the estimator in calculating the billed rate. |
| 19.23 / 2.70 | Form cav. in holl. wall, width: 50, a.b. & Cav. wall insulatn. fillg., thickness: 50, fibre glass. | As this item has been described previously, there is no need to repeat it in full. The insulation to the cavity is described and measured in m², stating the thickness as SMM P11.1.1.0.0. |
| 18.93 / 2.70 | Conc. blk. wall, thickness: 100, Thermalite insulatg. blks., b. & j. in g.m. (1:1:6). | Block walling measurements follow the rules laid down in SMM F10.1.1.1.0. Brick and block walling is deemed vertical unless described otherwise (SMM F10.D3). The slight variation in girth from the ¢ of cavity is not considered significant. |
| 19.23 / 0.08 | Bk. wall thickness: 215, comms. in Eng. bond in g.m. (1:1:6). (eaves fillg.) | |
| 2/ 2.70 | Bondg. end of ½B wall to xtg. bwk. & Bondg. end of 100 Thermalite conc. blk. wall to xtg. bwk. | Bonding of brick outer skin and Thermalite inner skin to the existing brick walls is measured in metres stating thickness of the wall in each case. It is assumed that the piers are built off the bases of the existing walls and so no foundations are measured. |

Piers
2.700
less rsj. 152
2.548

| | | |
|---|---|---|
| 0.33 / 2.55 | Isoltd. pier, thickness: 328, comms. in Eng. bond in c.m. (1:3). | Isolated piers are measured in m² stating the thickness with the classification as SMM F10.2.1.1.0. Facework is any work in bricks or blocks finished fair (SMM F10.D2). |

15.9

| | | |
|---|---|---|
| 2.55 | Proj., width: 215, depth: 102.5, comms. in Eng. bond in c.m. (1:3), inc. bondg. to conc. blkwk. (attchd. pier | The projecting pier is measured as a projection as a linear item, stating the width and depth of projection and including the bonding of the brick pier to the inner blockwork skin, as SMM F10.5.1.1.2. |

Int. ptns.

| | | |
|---|---|---|
| 2/ 2.45 / 2.70 | Conc. blk. wall, thickness: 100, in type B lightwt. agg. blks., size 440 x 215 to BS 2028, w. keyed fin., b. & j. in g.m. (1:1:6). | It is assumed that the partitions are built off the floor slab. A full description of the blocks is given in accordance with SMM F10.S1. No pointing is required as the wall surfaces will be plastered. The adjustment for the door opening will be taken when measuring the door. It is not necessary to take an item for bonding the partitions to the blockwork inner skin, as it is not another form of construction. |

Stlwk.

| | | |
|---|---|---|
| 1 | Stl. isoltd. struct. membr., plain beam, 152 x 89 x 17.09 kg/m, 2.45 m lg. to BS4 Pt 1, table 5, supptg. tbr. flat rf. | Billed in tonnes to classifications given in SMM G12.5.1.2.0, including use and details of construction. Building in ends of steel sections are not measured but if padstones were required they would form a separate enumerated item. |
| 1 | Ditto., 5.45 m lg. | |

Flr. Slab.
3.740
100
8.430
12.270
less holl. wall 255
12.015

Build up of length of floor slab in waste from figured dimensions on drawing.

| | | |
|---|---|---|
| 12.02 / 2.45 / 0.15 | Fillg. to make up levs., av. thickness ≤ 0.25m, obtnd. off site, broken bk. or st. | Hardcore filling is measured in m³ in accordance with SMM D20.10.1.3.1. |
| 12.02 / 2.45 | Compactg. fillg., blinded w. sand. | Compacting filling item as SMM D20.13.2.2.1, including the blinding material. |

15.10

| | | | | | | | |
|---|---|---|---|---|---|---|---|
| | | ALTERATIONS (Contd.) | | | | | |
| | 12·02<br>2·45<br>0·15 | In situ conc. bed (1:2:4/20 agg.), thickness ≤ 150.<br>less holl. wall  3·900<br>255<br>3·645 | Concrete beds are measured in m³ giving the thickness classification as SMM E10.4.1.0.0, and mix details as SMM E10.S1. | 8/ | 16·16 | Rf. membrs., flat. 50 × 200 treated sn. swd.<br>&<br>Indivdl. suppts., 50 × 40 (av.), diff. c.s. shapes, treated sn. swd.<br>add beargs.2/100  4·150<br>200<br>4·350 | Flat roof joists are classified as roof members as SMM G20.9.1.1.0.<br>Firring pieces are classified as individual supports and measured in metres, giving a dimensioned cross-section description as SMM G20.13.0.1.1 and G20.D7. |
| | 12·02<br>2·45<br>3·65<br>2·45 | Damp prfg., hor. Visqueen heavy duty polythene, 1000 gauge ld. on conc. (lounge extsn. | Horizontal damp-proof membrane measured as a superficial item, giving particulars of the material and nature of base on which it is laid (SMM J40.1.1.0.0 and J40.S1-2). | 8/ | 4·35 | Ddt. Rf. membrs., flat, 50 × 200 do. (hall extsn.<br>&<br>Add 50 × 200 do. | Lengths > 6·00 m in one continuous length are measured separately, stating the length. |
| | 3·65<br>2·45<br>3·74<br>2·45<br>4·51<br>2·45<br>0·90<br>0·10 | Scrd. to flr., lev., thickness: 38 ct. & sd. (1:3) stl. trowelled to rec. (lounge extsn.<br>PVC tiles on conc. (bar area<br>&<br>Plastic flr., width > 300, lev., 300 × 300 × 2 th. PVC tiles to BS 3261 Type A, fxd. w. adhesive on trowelled scrd. (m/s). (hall extsn. (bar area dr. opg.<br>2·450<br>900<br>215<br>3·565 | Screed is detailed in accordance with SMM M10.5.1 and M10.S1-6.<br>The description of the floor tiles includes the appropriate particulars listed in SMM M50.5.1.1.0 and M50.S1-8.<br>The floor finishings to the openings into the hall and lounge were included with the cutting opening items. | | 16·16 | Plate, 50 × 200 treated sn. swd. spld. fxd. to xtg. r.c. portal fr. (brr.<br>&<br>Indivdl. suppt., 50 × 50, irreg. shaped area, treated sn. swd. (tiltg. fillet<br>len.<br>16·170<br>less end beargs. 2/100  200<br>15·970<br>width<br>2·260<br>less 5/50  250<br>2·010 | The bearer is classified as a plate in accordance with SMM G20.8.0.1.0 and G20.D4, giving a dimensioned description.<br>The tilting fillet is classified as an individual support as SMM G20.13.0.1.4, giving a dimensioned overall cross-section description.<br>The length of the roof insulation is calculated and the width scaled from the large scale section, and the combined width of the joists deducted. |
| | 3·57<br>2·45 | Granolithic flrg., lev., thickness: 40, in 1 coat, 1 pt. ct. to 4 pts. granite chippgs. graded 6 to dust w. a. stl. trowelled fin. ld. on conc. (beer store | The granolithic paving description contains the appropriate particulars listed in SMM M10.5.1.1.0. Alternatively, the floor finishings could be measured along with wall and ceiling finishings. | | 15·97<br>2·01 | Insulatn. quilt, hor., fibreglass, 100 th., w. 100 laps, between rf. members @ 400 ccs. (mesd. o/a). | Insulation is measured the area covered in m² (SMM P10.M1), and giving the particulars listed in SMM P10.2.3.1.0 and P10.S1-3. The overall measurement allows for the quilt turned up against roof members. |
| | 0·90 | Dividg. strip 6 × 16 ebonite bedded in c.m. (1:3). (beer store/ bar area dr. opg. | Dividing strips are measured in metres in accordance with SMM M50.13.5.1.0, giving a dimensioned description. | | | Covergs. | |
| | | Flat Rf. Constn.<br>less<br>outer pt. of holl. 16·170<br>walls 2/155  310<br>15·860<br>add<br>laps at ptns. &<br>r.s.j. 3/100  300<br>16·160 | The overall lengths of the roof joists are calculated in waste, allowing for bearings.<br>A preamble clause will cover the details of the preservative treatment of all structural timber. | | 16·32<br>3·34 | Built up felt rf. coverg., flat pitch, to BS747 in 3 layers of asb. based felt, fully bondg. all layers in hot bit. compd. & coverg. w. 12 limestone chippgs. in bit. dressg. compd. ld. on t. & g. bdg. (m/s). | The length is taken from the previous dimensions for the asbestos cement sheeted roof and checked by scaling, while the width is scaled from the large scale section.<br>The roof covering is measured in m² stating the pitch as SMM J41.2.1.0.0 and particulars as SMM J41.S1-3. |

15.11    15.12

ALTERATIONS (Contd.)

| | | |
|---|---|---|
| | 16.320<br>3.340<br>900<br>20.560 | Build up of length of eaves in waste. |
| 20.56 | Built up felt eaves, gth. ≤ 200, in 2 layers w. mineral surfd. outer layer. | Built up felt at eaves is measured in metres stating the girth in stages of 200 mm as SMM J41.4.2.0.0. All labours are deemed to be included (SMM J41.C2). |
| 16.32 | Built up felt abut. gth. 200-400, in 2 layers w. mineral surfd. outer layer. | See SMM J41.3.2.0.0 for measurement of abutments. The girth includes dressing over the fillet and purlin.<br><br>Other types of flat roof covering are measured in 'Building Quantities Explained'. |
| 16.32<br>3.34 | Rf. bdg., width > 300, 25th treated sn. swd. t & g. nailed to swd. jsts. (m/s). | Roof boarding is measured in m² giving the width range and a dimensioned description (SMM K20.4.1.1.0). It is only classified as sloping if > 10° from the horizontal. All labours are deemed to be included (SMM K20.C1a).<br>The difference in length between the drip and the fascia is so minimal that it does not warrant adjustment. |
| | eaves | |
| 20.56 | Fascia bd. 25 x 200 treated wrot swd.<br>&<br>Indivdl. suppt. 19 x 38 treated sn. swd. (drip<br>&<br>Paintg. gen. isoltd. surfs. ext., wd., gth. ≤ 300, k.p.s. & ② u/c & ① gloss oil. | The fascia is a linear item as SMM G20.15.3.2.0, with a dimensioned overall cross-section description as is also the fillet (drip) as SMM G20.13.0.1.0 and G20.D7. Painting to fascia boards is described as general isolated surfaces and is measured in metres as ≤ 300 mm girth (SMM M60.1.0.2.0). External work must be so described (SMM M60.D1), and the nature of the base and painting system must be stated (SMM M60.S1-8). The sequence of the description is likely to appear unusual, but it follows closely the order of items in SMM7 and is logical and well suited to specification cross referencing and computerisation. |

15 · 13

| | | | |
|---|---|---|---|
| | | r.w. installtn.<br>20.560<br>add angles 2/100   200<br>20.760 | The gutter is measured on its centre line over all gutter fittings (SMM R10.M6). The description of the gutter is to include the type, nominal size, method of jointing and the type, spacing and method of fixing brackets (SMM R10.10.1.1.1). This type of gutter does not require painting. |
| | 20.76 | Gutter, st., nom. size: 100φ h.r. uPVC to BS 4576. w. union clip jts. & vinyl. bkts. at n.e. 900 ccs. scrd. to swd. | |
| 2/ | 1 | E.o. 100φ uPVC gutter for angle.<br>&<br>Ditto. for s.e.<br>&<br>Ditto. for nozzle outlet for 63φ downpipe, inc. cop. wire balloon gtg. | Then follow with enumerated fittings, measured as extra over the gutter (SMM R10.11.2.1.0). Note the use of a single item for a nozzle outlet incorporating a copper wire balloon grating, and method of grouping different items with similar quantities.<br>Balloon gratings can be enumerated in accordance with SMM R10.6.4.1.0 as a pipework ancillary. Alternatively, they can be included in the descriptions of the outlet items (SMM R10.M4). |
| 2/ | 3.10 | Pipe, st., nom. size: 63φ uPVC rwp w. push fit jts. & fxg. w. stand. vinyl pipe bkts. at n.e. 900 ccs. fxd. w. brass scrs. to plugs in bwk. | Downpipes are measured in metres, over all pipe fittings, and descriptions are to include type, nominal size, method of jointing and type, spacing and method of fixing supports and background as SMM R10.1.1.1.1. No swannecks are required. The jointing of the bottoms of the downpipes to the gullies is included in the gully items as SMM R12.10.1.1.0 and R12.D6 and C5. |
| | | Pitched Rf. to Hall | |
| 8/ | 16.07 | Rf. membrs, pitched 50 x 125 treated sn. swd. (purlins<br><br>len. of slope<br>3.500<br>Less<br>junctn. w<br>flat roof.   600<br>2.900 | All roof timbers except plates are classified as roof members, pitched, giving a dimensioned description (SMM G20.9.2.1.0). The lengths of the purlins have been calculated previously.<br><br>One slope is of shorter length than the other. |

15 · 14

## ALTERATIONS (Contd.)

| | | |
|---|---|---|
| 16.32<br>3.50<br>16.32<br>2.90 | Metal profiled rf. coverg., 'Trimaroof' composite stl. deckg. 55 th. supplied by H.H. Robertson & consistg. of ext. galvd. stl. sheets coated w. polyester resin, insulatn. core of isocyanurate foam, embossed, neoprene bonded & coated w. PVC. & fxg. to tbr. purlins @ 1100 ccs. w. galvd. drive scrs. & 76 × 76 × 1.6 th. plate washers @ 300 ccs., w. min. 150 side & end laps. | Roof decking descriptions shall include the particulars listed in SMM H31.S1-5. Roof coverings are measured in m² stating the pitch as SMM H31.1.1.0.0. |
| 16.32 | Propy. ridge piece to ditto., 300 ⌀. | Flashings, eaves, ridges and the like are each given separately in metres, with a dimensioned cross-section description. |

To take
Gutters & downpipe finishings, doors.

It will be necessary to replace or refix the gutters and one downpipe to the pitched roof and this will constitute an architect's query.

Note: The structure items will be followed by finishings, doors and the adjustment of door openings. These have not been taken off as sufficient examples of wall and ceiling finishings have already been provided in Example 12. Two examples of doors accompanied by three window examples, together with the adjustment of openings are included in Building Quantities Explained. Limitations of space in this book create the need to avoid repetition and also help to keep book production costs to a reasonable level in these inflationary times.

### Drainage
### Gully connectns.

```
              1.260
              1.040
           2) 2.300
              1.150
              
              300
existing 150⌀ drain
invt. (averaged)   1.150
                2) 1.450
av. depth of pipe   725
add conc. bed      150
av. o/a depth     875
```

Calculations of the average depth of the new branch pipe trenches, taking the pipes as 300 mm deep at the head of each branch, and allowing for a concrete bed as part of the pipe protection rendered necessary by their shallow depth.

15.15

| | | |
|---|---|---|
| 2/ | 1.50 | Exc. tr. for pipe ≤ 200 nom. size, av. depth of tr. ≤ 1.00 m.<br>&<br>Pipes in trs. 100 BS clay to BS 65 Pt.2 jtd. w. flex. mech. jts.<br>&<br>In situ conc. bed & surrd. (1:3:6/20 agg.), width of bed : 700 & thickness of bed & surrd : 150, nom. size of pipe : 100. |
| 2/ | 1 | Pipe accessory : gully, clay trapped to BS 539 w. 100 ⌀ outlet, 75 back inlet & 150 × 150 sq. c.i. galvd. gratg. |
| 2/ | 2 | E.o. 100 clay pipe for bend.   (gullies |

Trenches for pipes ≤ 200 mm nominal size are grouped together (SMM R12.1.1.2.0). The depth is given in average depth stages of 250 mm. The kind of pipe does not have to be specifically mentioned in pipe trench excavation descriptions.
The description of the pipes is to include the kind of material (clay), quality of pipe (BS to BS 65 Part 2), nominal size (100 mm) and method of jointing (flexible mechanical joints) as SMM R12.8.1.1.0 and R12.S1 and S4.
Iron pipe in runs ≤ 3.00 m in length are so described, stating the number, as short lengths are expensive.
Beds and surrounds to pipes are measured in metres, stating the width of the bed and thickness of bed and surround and materials of which constructed, and the nominal size of pipe (SMM R12.6.1.1.0). Any formwork required is deemed to be included (SMM R12.C2).

Gullies are enumerated, giving a dimensioned description as SMM R12.1D.1.1.0. Jointing to pipes and bedding in concrete are deemed to be included (SMM R12.C5). The dimensions stated for accessories are to include the nominal size of each inlet and outlet (SMM R12.D7).

Bends are enumerated as extra over pipes (SMM R12.9.1.1.0), and cutting and jointing pipes are deemed to be included (SMM R12.C4). Two bends are taken to each square gully (one for direction and the other for gradient).

15.16

ALTERATIONS (Contd.)

| | | | |
|---|---|---|---|
| 2/ | 1 | Exc. to expose xtg. 150 φ clay pipe ov. 1·15 m dp., provide & insert 150 x 100 branch pipe, joint to xtg. pipe w. double collar & renew one len. of 150 φ pipe. | The connection of the branch pipes from the new gullies to the existing 150 mm drain pipe is enumerated, describing all the work involved. Alternatively, a saddle connection could be formed. |
| | Item | Allow for testg. & commsng. drainage installtn. after backfillg. w. an air test whereby a pressure equiv. to 100 hd. of water shall not fall to less than 75 hd. after 5 mins. | Item stating the method of testing as SMM R12.17.1.0.0. A more comprehensive example of the measurement of drainage work, including manholes, is given in 'Building Quantities Explained'. |
| | Item | Allow for disposal of surf. water. | Item for keeping excavations free of surface water as SMM R12.3.1.0.0. Where excavation is measured below groundwater level, an item is included for disposal of groundwater (SMM R12.3.2.0.0). |

15·17

# BIBLIOGRAPHY

Royal Institution of Chartered Surveyors and Building Employers Confederation. *Standard Method of Measurement of Building Works: Seventh Edition (SMM7)* (1988)

Royal Institution of Chartered Surveyors and Building Employers Confederation. *Bills of Quantities: A Code of Procedure for Building Works* (1988)

Building Project Information Committee. *Common Arrangement of Work Sections for Building Works* (1987)

Building Project Information Committee. *Project Specification: A Code of Procedure for Building Works* (1987)

Building Project Information Committee. *Production Drawings: A Code of Procedure for Building Works* (1987)

Co-ordinating Committee for Project Information. *Co-ordinated Project Information for Building Works, a guide with examples* (1987)

Property Services Agency, Royal Institution of Chartered Surveyors and Building Employers Confederation. *SMM7 Library of Standard Descriptions* (1988)

Fletcher, Moore, Monk and Dunstone. *Shorter Bills of Quantities: The Concise Standard Phraseology and Library of Descriptions*. Builder Group (1986)

I. H. Seeley. *Building Quantities Explained*. Fourth Edition. Macmillan (1988)

A. Smith. *Computers and Quantity Surveyors*. Macmillan (1989)

I. H. Seeley. *Quantity Surveying Practice*. Macmillan (1984)

I. H. Seeley. *Building Economics*. Third Edition. Macmillan (1983)

Chartered Institute of Building. *Code of Estimating Practice* (1984)

Institution of Civil Engineers and Federation of Civil Engineering Contractors. *Civil Engineering Standard Method of Measurement: Second Edition*. Telford (1985)

I. H. Seeley. *Civil Engineering Quantities*. Fourth Edition. Macmillan (1987)

Joint Contracts Tribunal. *Standard Form of Building Contract with Quantities* (1980)

The Aqua Group. *Pre-contract Practice for Architects and Quantity Surveyors*. Collins (1986)

Society of Chief Quantity Surveyors in Local Government. *The Presentation and Format of Standard Preliminaries for use with JCT Form of Building Contract with Quantities 1980 Edition* (1981)

Greater London Council. *Preambles to Bills of Quantities*. Architectural Press (1980)

Royal Institution of Chartered Surveyors. *Definition of Prime Cost of Daywork carried out under a Building Contract* (1981)

Royal Institution of Chartered Surveyors. *Schedule of Basic Plant Charges for use in connection with Dayworks under a Building Contract* (1981)

Royal Institution of Chartered Surveyors, Quantity Surveyors Division. *The Future Role of the Chartered Quantity Surveyor* (1983)

L. Black (Editor). *Builders' Reference Book*. Twelfth Edition. Northwood Books (1985)

Chartered Institute of Building Services. *Institution of Heating and Ventilating Engineer's Guide* (1981)

Heating and Ventilating Contractors' Association. *Specification for Sheet Metal Ductwork (Low and High Velocity)* (1977)

Institution of Electrical Engineers. *IEE Regulations for Electrical Installations: Regulations for the Electrical Equipment of Buildings* (1981)

# INDEX

Abbreviations 22
Accuracy 26, 148
Adjustments for openings 148
Air turns 133
Alterations
   bonding new work to old 146, 153, 154, 156, 157
   cutting openings 146, 147, 153, 154
   filling openings 146, 153
   general procedure 146–8
   inserting lintels 147
   making good 147
   preparing tops of walls 153, 154
   removal of fittings 147, 154
   sequence of measurement 147–8
   temporary screens 147
   temporary work 146–7, 155
   wedging and pinning 46, 146, 153, 154
Anchorages 67
Approximate quantities 26
Arches 48, 49, 50, 59–60
Architraves 109
Ashlar 50, 62
Asphalt work
   preamble clause heading 14
   special requirements 150
Automated measurement 25

Backings to wall linings 112, 113, 114, 117
Balustrading 61, 78
Bands to brickwork 47–8, 54, 56
Bar schedule 64, 73, 150
Beams
   reinforced concrete 63, 65
   steel 69
Bedding frames 97, 100
Beds to floors 37, 63, 150, 158
Benching 104–6
Bibliography 162
Bill diagrams 129, 131, 133, 151
Bill production 23–5
Bitumen felt roofing 158–9
Blockwork
   measurement 47, 157

preamble clause headings 13
Boarded ceiling 112
Bolts 84, 151
Bonding new work to old 146, 153, 154, 156, 157
Bored cast in place piles 38–40
Boreholes 27
Brickwork
   arches 48
   bands 47–8, 54, 56
   bedding frames 97, 100
   boiler seatings 48
   bonding new work to old 146, 153, 154, 156
   cavity formation 156, 157
   cavity trays 48
   chimney shaft 48, 53–7
   damp-proof courses 36, 37, 48, 55, 147, 156, 158
   facework 36, 47–8, 54, 55, 57, 150, 154, 156, 157
   facework ornamental bands 47–8, 54, 56
   facework projections 56, 57
   firebrick lining 55, 57
   firebrick paving 55
   flue linings 48
   general brickwork 47, 156, 157
   hollow walls 156, 157
   piers 157
   plinth cappings 55, 56
   preparing tops of walls 153, 154
   projections 35, 53–4, 56, 57, 157
   rake out joints 77, 116
   sills 57
   substructure 35
   walling 35, 46
   wedging and pinning 46, 146, 153, 154
Bridge 80–2
Builder's work 124, 144, 145, 151

Cables 135, 136, 139, 141, 142
Carpentry
   members 91–2, 150, 158, 159
   preamble clause headings 13
Carpeting — supply 11

Casings 92
Cast in place piles 29, 38–40
Cavity fill
   concrete 156
   insulation 157
Cavity formation 156, 157
Ceramic tiles
   to floors 108, 120
   to walls 108, 117
Chemical damp-proof courses 147
Chimney shaft 52–7
Cleaning 8
Cleats 84, 88, 89
Coach bolts 97
Code of Procedure for Measurement of Building Works 20, 26, 49, 137, 138
Code of Procedure for Single Stage Tendering 1
Coding systems 24–5
Columns 74
Common Arrangement 12, 19–20, 21, 123
Compacting bottom of excavation 28, 34, 40, 46, 53, 155
Compacting filling 36, 54
Component drawings 26, 78, 94, 97
Composite
   decking 66, 71
   items 92–4, 103, 150
Computerised billing systems 23–4
Computers 22–5
Concrete
   backing to brick wall 46
   beam casings 65
   beams 63, 65
   beds 37, 55, 63, 82, 150, 158
   bridge 80–2
   casing to steel beam 71
   cavity fill 156
   column bases 63
   columns 74
   construction joints 82
   contractor designed work 71
   floors 65, 68–71
   formwork 34, 40, 46, 64–5, 69–70, 71, 74, 75, 76, 150

Concrete (cont'd)
  foundations  34, 40, 46, 53, 155
  hack surface  77, 107
  *in situ*  63–4
  lintels  57
  piles  38–41
  pipe protection  160
  preamble clause headings  12–13
  precast work  57, 63, 66, 69, 71, 81–2
  prestressed work  63, 66–7
  reinforced  63, 69, 74
  reinforcement  39, 40, 64, 65, 70–1, 74, 75–6, 77, 150
  roofs  65
  slabs  63, 65, 69, 74
  staircase  65–6, 72–8
  surface treatment  37, 40, 55, 64, 76, 77, 82
  upstands  65
Conduit  135, 136, 141, 142
Construction joints  82
Contingencies  11
Contract  3–4
  appendix to conditions  4
  schedule of clause headings  3–4
Contractor designed work  71
Contractor's general costs  7–10
Contractor's liability  4
Contractor's staff  7
Copings  57
Corner piles  42
Cornices  61, 108, 109, 115
Cultivation  17
Cutting openings  146, 147, 153, 154

Dado rails  109, 115
Damp-proof courses  36, 37, 48, 55, 147, 156, 158
Damp-proof membranes  37, 158
Dayworks  11–12
Decorative papers  109
Demolition
  existing buildings  146–7, 153
  measurement procedures  146–7, 148
  preamble clause headings  12
  salvageable materials  6, 146
Descriptions
  framing  22, 26
  of works  3
Digitiser  25
Dimensioned diagrams  26, 76, 93, 97, 151, 168
Direct works  11
Disbursements arising from employment of workpeople  9

Disposal of excavated material  28, 33, 39, 53, 155
Disposal of groundwater  46
Disposal of surface water  28, 35, 46, 156, 161
Disposal systems  123
Distribution boards  137, 138, 142
Dividing strips  118–19, 158
Documents
  confidentiality  5
  upkeep  5
Door frames  97
Doors  14, 93–4, 97
Dowels  50
Drainage
  clay pipes  160
  concrete protection  160
  excavate pipe trench  160
  excavate to expose existing drain  161
  gullies  160
  preamble clause headings  15–16
  testing  161
Drawings
  'as fitted'  134, 138
  'as installed'  124
  bill diagrams  129, 131, 133, 151
  bill preparation  2
  component  26, 94, 97, 103
  contract  2
  dimensioned diagrams  26, 76, 93, 97, 108, 151
  inspection of  2
  location  29, 47, 63, 83
  mechanical services  121, 122, 123, 124, 128
  tender  2
Ducting  124, 126, 132–3, 134
  brackets  133, 134
  fittings  132, 133, 134

Earthwork support
  left in  46
  measurement  27–8, 34, 40, 45, 46, 53, 155
  preamble clause  17
Ebonite dividing strips  118–19, 158
Electrical services
  accessories  137, 142, 143, 144
  'as fitted' drawings  138, 145
  builder's work  144, 145, 151
  cable tray  136
  cable trunking  136
  cables  135, 136, 139, 141, 142
  conduit  135, 136, 141, 142
  cutting or forming holes, etc.  145
  description of installation  141
  detailed measurement  135–8
  distribution boards  137, 138, 142

distribution sheet  140
drawings  135
final circuits  136–7, 143, 151
fittings  135, 136, 138, 139
fixed apparatus circuit  138
gangs  137, 144
identification labels  137, 138, 139, 144
IEE Regulations  135, 141
installation  11
junction boxes  136
lamps  137, 144
light fittings  137, 139
lighting installation  138, 143–4
luminaires  137, 143–4
mains installation  138, 141–3
mark position of holes  138, 142, 143, 144
measurement procedures  135, 151
notation  136
pendants  137, 139, 143
points  137
prime cost sum  11
socket outlets  137, 139, 142, 143
specification  138–9
sundries  138, 142, 143, 144, 145
switches  137, 139, 144
switchgear  137
testing  138, 142, 143, 144
Electronic measurement  25
Employer's liability  4
Employer's requirements  4–7
Examination approach  149
Examination techniques  148–9
Examiners' reports  148–51
Excavated material
  disposal  17, 28, 39, 41, 53
  re-use  16
Excavation
  adjustment  34, 35, 46, 53, 54
  below required levels  17
  breaking out existing materials  27
  classifications  27
  compacting bottom of excavation  28, 34, 40, 46, 53, 155
  disposal of excavated material  17, 28, 39, 41, 53, 155
  disposal of water  28, 35, 46, 156
  earthwork support  17, 27–8, 34, 40, 45, 46, 53, 155, 161
  filling  17, 28, 33, 34, 36, 45, 53, 155, 156, 157
  grass seeding  17
  groundwater  16, 27, 28, 46
  hardcore  16
  levels  16
  measurement procedure  27–9, 149–50

Excavation (cont'd)
  pile caps   39
  pipe trench   160
  pits   53
  preamble clause headings   12
  re-use of excavated material   16
  site preparation   16, 27
  sloping sites   28–9, 150
  stepped foundations   28–9
  surface treatment   28, 34, 40, 46, 53, 155
  topsoil   16–17, 33, 53, 150, 155
  trenches   27, 33, 34, 45, 149–50, 155
  trial holes   16
  typical preamble clauses   16–17
  unauthorised   16
  weaknesses in measurement   149–50
  working space   27

Facework   36, 47–8, 54, 55, 56, 57, 150, 154, 156
Fans   8, 133
Fascia board   159
Fees and charges   11
Fencing
  preamble clause headings   15
  temporary   7, 9
Fertilising   17
Fibreboard acoustic ceiling   113
Fibreglass insulation   158
Fibrous plasterwork   109
Filler piece   97
Filling
  compacting   36, 157
  excavations   17, 28, 33, 34, 36, 45, 53, 54, 155, 156, 157
  openings   146, 153, 156
  preamble clauses   16
Final circuits   136–7, 143
Finishings
  backings   112, 113, 114, 117
  beams   107
  beds   108
  ceilings   107–8, 112–13
  ceramic tiles   108, 117, 120
  cork tiles   108, 120
  cornices   108, 109, 115
  curved work   107
  fibreboard acoustic tiles   113
  fibrous plaster   109
  floors   108, 118–20
  granolithic paving   77, 158
  measurement principles   107–8
  measurement sequence   107
  plaster   107–8, 112, 113, 114, 115, 116, 154
  plasterboard   107

  plywood panelling   115
  preamble clause headings   14–15
  PVC tiles   109, 119
  quarry tiles   108, 119, 151
  renderings   108
  rounded angles   108, 112
  schedule   107, 111
  screeds   108, 118, 119
  skirtings   117–18
  sprayed mineral fibre coatings   109
  staircase areas   108
  terrazzo *in situ*   113, 114, 118
  terrazzo tiles   108
  Tyrolean finish   114
  walls   108, 113–17
  wood blocks   108, 119
Firebrick
  lining   55, 57
  paving   55
Firring pieces   158
Fitments   92–3
Flooring
  ceramic tiles   108, 120
  granolithic paving   77, 158
  PVC tiles   119, 158
  quarry tiles   108, 119
  terrazzo paving   118
  timber   92
  wood blocks   108, 119
Formwork
  beam casings   71
  beams   70, 75
  columns   74
  foundation sides   34, 40, 46
  mortices   78
  slab edges   70
  slab soffits   69, 75
  upstands   76
Foundations   27–9, 33–6, 40, 46, 53, 155
Frames
  steel   86–90
  wood   91–2, 93
Furniture   6

General attendance   8, 9, 10
General description of works   2
General Summary   12
Glazed screens   93–4, 99–100
Glazing
  beads   93, 94, 98, 99
  doors   98
  lights   93, 99
  preamble clause headings   14
  screens   94
Granolithic paving   77, 158
Granolithic skirting   77, 109
Granolithic steps   77–8
Grass seeding, preamble clauses   17

Greater London Council
  preambles   12, 16
Grillages   88
Ground conditions   16
Grounds   115, 117, 118
Groundwater
  measurement   27, 28, 46
  preamble clause   16
Gullies   160
Gutters   159

Hack surface of concrete   77, 107
Handrails   78
Hardcore
  measurement   36, 54, 157
  preamble clause   16
Hardstanding   9
Head   97
Health and Safety at Work Act 1974   147
Hoardings   7, 9
Hollow block suspended construction   71
Hollow walls   156, 157
Humidity   7, 11

Identification discs/labels   131, 132, 137, 138, 139, 144
IEE Regulations   135, 141
IHVE Guide   122
Individual supports   113, 115, 158, 159
Information technology   22
Inlet louvres   126, 132
Insulation
  cavity   157
  pipes   126, 134
  roof   158
Insurances   11
Ironmongery   93, 98
Isolated structural steel members   83, 151, 157

Jamb stones   48, 59
Jambs   97
JCT Standard Form of Building Contract with Quantities   1, 3–4
Junction piles   42

Kickers   65
Kneebraces   89

Lamps   137, 144
Leather covering to benchtop   106
Levels, preamble clause   16
Library of Standard Descriptions   21, 24
Light fittings   137, 139
Lighting and power for the works   8

Lighting installation   138, 143–4
Linings   92, 104, 105, 112, 115
Lintels   57, 154
Local authority fees and charges   4
Location drawings   29, 47, 73, 83
Logical approach to
    measurement   149
Louvres   126, 132
Luminaires   137, 143–4

Mains installation   138, 141–3
Maintain
    adjoining buildings   5
    live services   5
    roads   5, 8–9
Making good   147
Mark positions of holes   132, 134,
    138, 142, 143, 144
Masonry
    arches   48, 49
    ashlar   50, 62
    balustrade   61
    capping   61
    columns   48
    compression joint   62
    copings   48, 50, 61
    cornices   61
    cutting   48
    description   49
    jambs   48, 59
    labours   48
    measurement   48–50
    preamble clause headings   13
    slabbing   50
    special purpose   59–60
    steps   48, 59
    surround to door opening   48,
        59–60
    voussoirs   60
Measurement
    accuracy   26
    descriptions   22, 26
    dimensioned diagrams   26, 76, 93,
        97, 108, 151
    examination approach   149
    examining drawings   26
    presentation   26–7
    procedures   26–7
    sequence   26
    subheadings   26
    use of figured dimensions   26
    waste   26
Mechanical services
    ancillaries   122, 123
    approximations   121
    builder's work   124
    co-ordination   121–2
    cutting holes   123, 130
    dampers   126
    description of building   125

description of installation   129
drawings   121, 122, 123, 124, 128,
    134
ducting   124, 126, 132–3, 134
equipment   123–4
fan convector   125, 132, 133
fittings   122, 123, 126, 130, 131
grilles   126
identification discs   131, 132
inlet louvres   126, 132
insulation   126, 134
loose keys   134
mark position of holes   132, 134
measurement approach   122
measurement procedures   121–2
pipe sleeves   122, 125, 130
pipework   122, 123, 125, 126,
    130, 131
plant   8, 9, 122
reference information   122
specification   124–6
statutory requirements   129
supports   123, 124, 125, 126, 131,
    133, 134
systems   123
testing   132, 134
valves   122, 123, 126, 131
Metal mesh lathing   109
Metalwork
    balustrading   78
    base plates   104
    bench base   103
    composite items   92–4
    fascia plate   105
    framework   104
    painting   57, 78, 89–90
    pedestals   103
    soot door   57
    steel roof decking   160
Microcomputers   22, 23
Mortices   78

Name boards   7
National Building Specification   19
National Engineering Specification   19
Noise control   5, 6
Nominated sub-contractors   10
Nominated suppliers   10

Off site soil disposal   17, 45
On site soil disposal   33, 39, 41, 53,
    155
Operation of services   7
Operatives' disbursements   9
Overheads   11, 12

Padstones   69
Painting and decorating
    ceiling plaster   109, 112, 113
    cornices   109, 116

decorative papers   109
doors   94, 99
glazed screens   100
handrails   78
metalwork   57, 78, 89–90
multi-coloured work   94, 107, 109
plant rooms   109
preamble clause headings   15
screens   94, 99, 100
services   130, 145
skirtings   118
staircase areas   109
wall plaster   109, 114, 115, 116
wax polish block flooring   108, 119
wood ceiling   112
woodwork   78, 94, 99, 100, 112,
    115, 118, 159
Panelling   115
Phraseology Independent Billing
    System   25
Picture rails   109
Piers   157
Piles
    caps   39–40
    cast in place   29, 38–40
    corner   42
    cutting off tops   39, 41
    enlarging bases   39
    junction   42
    location drawings   29
    measurement procedure   29–30
    preamble clause headings   12
    preformed concrete   29, 38, 41
    rig standing time   40, 41, 42
    steel   30, 38, 42
    testing   40, 41, 42
Piped supply systems   123
Pipework   122, 123, 125, 126, 130,
    131
    preamble clause headings   16
    supports   123, 125, 130, 131
Plant   8, 122
Plant charges   12
Plant rooms   107, 109, 122, 132, 151
Plaster
    ceilings   107–8, 112, 113
    preamble clause headings   14–15
    rounded angles   108, 112, 114,
        115, 116
    walls   108, 114, 115, 116, 154
Plates
    metal   84, 88, 89, 125, 130
    wood   92, 158
Plywood panelling   115
Post-tensioning   66–7, 81–2
Preambles   12–19
    clause headings   12–16
    specimen clauses   16–19
Precast concrete   57, 63, 66, 81–2,
    154

Preformed concrete piles  29, 38, 41
Preliminaries  1–10
  contract  3–4
  contractor's general costs  7–10
  contractual aspects  3
  employer's requirements  4–7
  introductory information  1–2
  nominated sub-contractors  10
  nominated suppliers  10
  preliminary information  2–3
Prepare tops of walls  153, 154
Presentation  26–7, 148
Prestressed concrete
  anchorages  67
  bridge  80–2
  bridge units  81–2
  construction joints  82
  ducts  67
  measurement  66–7
  tendons  67
Pre-tensioning  66–7
Prime cost items  10, 11, 12, 47, 83, 93, 97
Production drawings  20, 22
Profit  10, 11, 12, 47, 93, 97
Programming  6
Project particulars  2
Project specification  20, 22
Projections to brickwork  35, 53–4, 56, 57, 157
Protection of the works  8
Protective clothing  146–7
Provisional sums  10, 11, 26
Purlins  83, 151, 159
PVC tiles  119, 158

Quality control  5
Quarry tiles  108, 119, 151, 154
Query list  26, 77, 121, 149

Rainwater goods, preamble clause headings  15
Rainwater pipes  159
Rake out joints of brickwork for key  77, 116
Refix windows  154
Removal of fittings  147, 154
Removal of roofs  155
Removing rubbish  8
Removing stains  147
Renovations  147
Repairs  147
Replacement of topsoil  36
Report on the Future Role of the Chartered Quantity Surveyor  22
Revision notes  148–51
Roof trusses  90
Roofing
  asphalt  150
  bitumen felt  158–9

  boarding  92
  composite steel decking  160
  flat  92, 158
  gutter  159
  pitched  91–2, 150, 159–60
  preamble clause headings  13–14
  rainwater pipes  159
  timbers  91–2, 158, 159
  trusses  90
Rubbish disposal  8, 11
Rubble walling  50

Safety, health and welfare  5–6, 8
Salvageable materials  6, 146, 153
Scaffolding  9
Schedules  73, 122, 128, 148
  of finishings  107, 111
Screeds  108, 118, 119, 120, 154, 158
Screens  93–4, 99
Security  6, 8
Sequence of measurement  26
Shoring  146–7
Shorter Bills of Quantities  21
Sidelights  98
Sills  97, 154
Site
  access  6
  accommodation  8
  administration  8
  attendance  9
  description  2–3
  preparation  16, 27
  use of  6
Skirtings  109
  granolithic  77, 109
  terrazzo  109, 117
  wood  109, 117, 118
Slabs  63, 65, 69, 74
Sloping site measurement  28–9, 150
Small plant and tools  9
SMM7  20, 21–2, 26, 149
Society of Chief Quantity Surveyors in Local Government  1, 3, 11
Socket outlets  137, 139, 142, 143
Soot door  57
Special attendance  10
Specification  20, 22
  electrical  138–9
  mechanical  124–6
Staircases  109
  areas  107, 151
  reinforced concrete  65–6, 72–8
  timber  94
Stanchions  88–9
  bases  88
Standard classifications  20–1
Statutory authorities  10
Statutory regulations  129
Steel balustrade  78

Steel piles  30, 38, 42
Steel reinforcement
  bars  39, 64, 70–1, 74, 75–6, 77, 150
  fabric  40, 64
  schedule  64, 73, 150
Steelwork tables  84
Stepped foundations  28–9, 32–7
Stone
  arches  48, 50, 59–60
  ashlar  50, 62
  balustrade  61
  capping  61
  cast  50
  columns  48
  compression joint  62
  copings  48, 50, 61
  cornices  61
  description  49
  jambs  48, 59
  rubble walling  50
  slabbing  50
  special purpose  59–60
  steps  48, 59
  surround to door opening  48, 59–60
  voussoirs  60
  walls  48
Structural steel framing
  beams  69, 89
  bolts  84
  condition  18
  connections  19, 69, 84
  dimensions  18
  erection  18, 69, 83, 85
  fabrication  17, 19, 69, 83–4
  grillages  88
  gusset plates  84, 88
  holding down bolts  83
  joints  19
  kneebraces  89
  marking  18
  materials  18
  measurement principles  83–4
  measurement procedure  84–5
  metallic coatings  18, 84
  painting  18, 19, 89–90
  particulars  83–4
  plates  84, 88, 89
  preamble clause headings  13
  roof trusses  90
  shop details  17
  stanchions  88–9
  standard of construction  17
  surface treatment  84, 85
  tests  18
  typical preamble clauses  17–19
  weight  18, 84
  workmanship  18–19
Sub-letting  5

Substructures, measurement
    procedures   26–7
Supplementary tables (SMM7)   21–2
Surface finishes   14
Surface treatment
    excavation   28, 34, 40, 46, 53, 155
    concrete   37, 40, 55, 64, 76, 77, 82
    steel   84, 85
Surface water disposal   28, 35
Switches   137, 139, 144
Symbols   22
Systematic approach to
    measurement   149

Take off list   77, 153
Telephones   7, 8
Temperature   7, 11
Temporary accommodation   6–7
Temporary fans   9
Temporary fencing   7, 9
Temporary hoardings   7, 9
Temporary roads   8
Temporary roofs   147, 155
Temporary screens   7, 147
Temporary telephones   7, 8
Temporary works   8, 9
Tenders   1–2, 4
Tendons   67
Terrazzo
    paving   118
    rounded angles   114
    skirtings   117
    wall linings/tiles   103, 113, 114
Testing
    drain pipes   161
    electrical lighting installation   144
    electrical mains installation   142
    electrical power installation   143
    hot water heating installation   132
    materials   5
    piles   40, 41, 42

provisional sum   11
    steelwork   18
    warm air heating and extract
        installation   134
Tile finishes to walls and floors   15
Timber particulars   91
Timesing   148
Topsoil
    measurement   33, 53, 150, 155
    preamble clause   16–17
Toxic materials   146–7
Traffic regulations   9–10
Transom   97
Trench excavation   27, 33, 34, 45,
        149–50, 155
Trial pits   16, 27, 44
Tyrolean finish   114

Underpinning
    cutting away existing
        foundations   30, 45
    measurement   30–1, 44–6
    pinning up new work   30, 46
    preliminary trenches   30, 45
    temporary supports   30, 45
    underpinning pits   30, 45–6
    width allowances   30

Valve identification discs   131, 132
Valves   122, 123, 126, 131
Ventilation systems   123

Waste   26, 108
Water for the works   8
Waterproof membranes   36, 158
Wax polish block flooring   108, 119
Weather bars   98
Wedging and pinning   46, 146, 153,
        154
Windows   11, 14, 94
Winter building   7

Wood block flooring   108, 119
Woodwool slab decking   13
Woodwork
    benching   103–6
    benchtop   104–6
    boarded ceiling   112
    carpentry timbers   91–2, 150, 158,
        159
    casings   92
    composite items   92–4
    doors   14, 93–4, 97
    firring pieces   92
    first fixings   91–2, 150, 158, 159
    fitments   92–3
    flooring   92
    frames   97
    glazed screens   94
    handrail   78
    individual supports   113, 115, 158,
        159
    ironmongery   93, 98
    leather covering   106
    linings   92, 104, 105, 112
    painting   78, 94, 99, 100, 112,
        115, 118, 159
    panels   105
    preamble clause headings   13
    roof boarding   92, 159
    roof timbers   91–2, 150, 158, 159
    screens   94, 100
    sidelights   98–9
    staircases   94
    timber framing   91–2
    timber particulars   91
    windows   11, 14, 94
Working hours   6
Working space   27
Works
    management   5, 7
    programme   6
    services   8
    temporary   8